1000 PACKAGE DESIGNS

ROCKPORT

First published in the United States of America by
Rockport Publishers, a member of
Quayside Publishing Group
100 Cummings Center
Suite 406-L
Beverly, Massachusetts 01915-6101
Telephone: (978) 282-9590
Fax: (978) 283-2742
www.rockpub.com

ISBN-13: 978-1-59253-445-6
ISBN-10: 1-59253-445-7

10 9 8 7 6 5 4 3 2 1

Design: Grip Design
Art direction: Kelly Kaminski, Kevin McConkey
Cover and interior design: Joshua Blaylock, Nikki Lo Bue

Printed in China

1000 PACKAGE DESIGNS

a comprehensive guide to packing it in

GRIP ○ CHICAGO

ROCKPORT PUBLISHERS

BEVERLY MASSACHUSETTS

THE TABLE

introduction and about the author
{ BEAUTY }
6-7

awesome!
{ MODERN }
8-63

gasp!
{ PLAYFUL }
64-115

word!
{ URBAN }
116-147

Grip

Grip is a marketing and design firm dedicated to high-value corporate communication for professional services and luxury brands. Companies turn to Grip when they need an approach to marketing that is rooted in research and strategy. Servicing clients in a multitude of industries, we provide the most appropriate voice for any communication vehicle. Our clients have labeled us the "thinking person's agency," a title we live up to with every assignment.

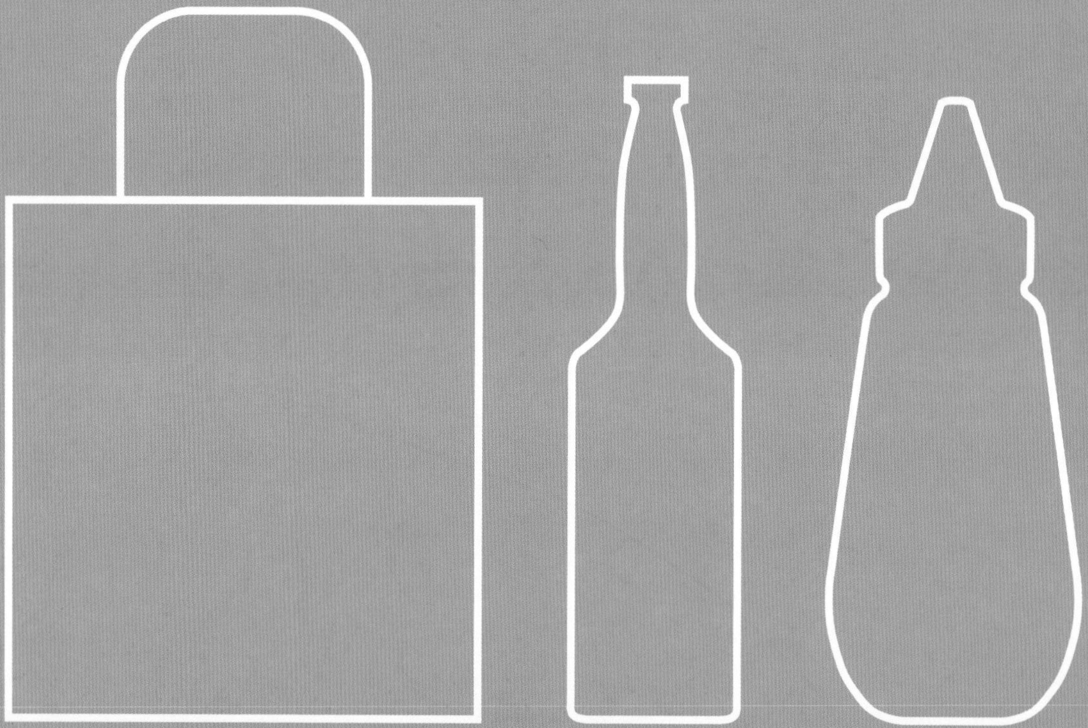

Have you ever bought something based solely on the merit of its package?

So have we. In our bathrooms, kitchens, and offices sit lotions we never tried, oils we never tasted, and pens we never . . . well, you get the idea. Each of these products were purchased irrespective of the merits of their contents. There may be better performing hand soap somewhere, but to grace the countertop of our bathroom requires a little design savvy. Appreciating form over function is in the DNA of most designers, and to you, the aesthetically inclined, we dedicate this book of beautiful, creative, innovative packages.

Where does one draw the line between package and product? In the rapidly evolving world of consumer marketing, trends like sustainability often call for less packaging. We believe that in the less-is-more greening of package design lies a beautiful harmony of product and package. It is with this in mind that some of the packaging in this book may seem more about the product but after all, if the product is good, why should the package stand in the way of immediate appreciation?

Organized by style, rather than product or industry type, the following pages are meant to inspire, provide a landscape of trends, and to prove once and for all that sometimes we all judge a product by its package.

Our selection process was filled with ecstatic moments of agreement and outright contention as we assembled a collection that represents the state of the industry. The winning selections were broken down into the following groupings: modern, playful, urban, classic, functional, and vintage. All in all there were many more entries than we had room to fit into this book and we would like to send our sincerest thanks to everyone who graced us with their time and design. Every day the process of curating the collection was exciting and surprising as we shamelessly indulged our love affair with beauty and the box.

awesome!

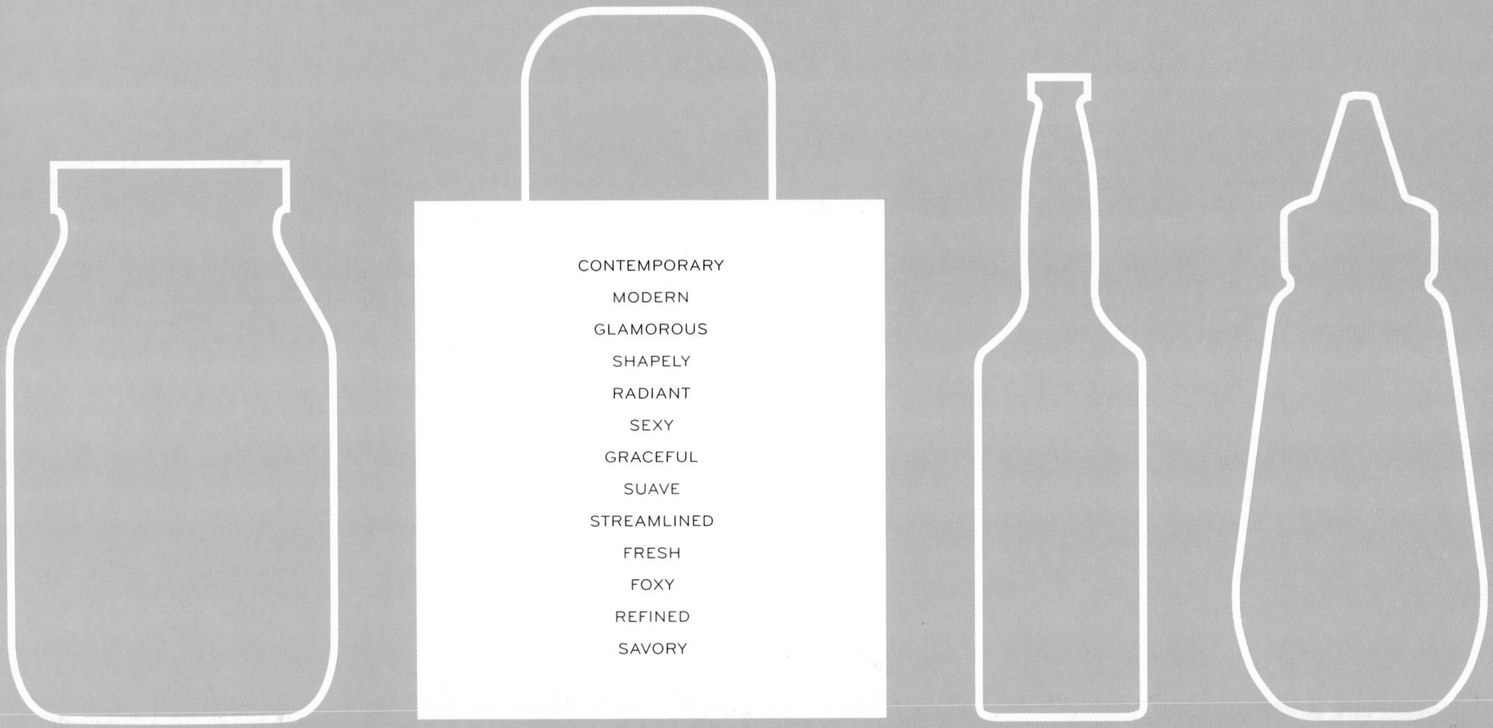
CONTEMPORARY
MODERN
GLAMOROUS
SHAPELY
RADIANT
SEXY
GRACEFUL
SUAVE
STREAMLINED
FRESH
FOXY
REFINED
SAVORY

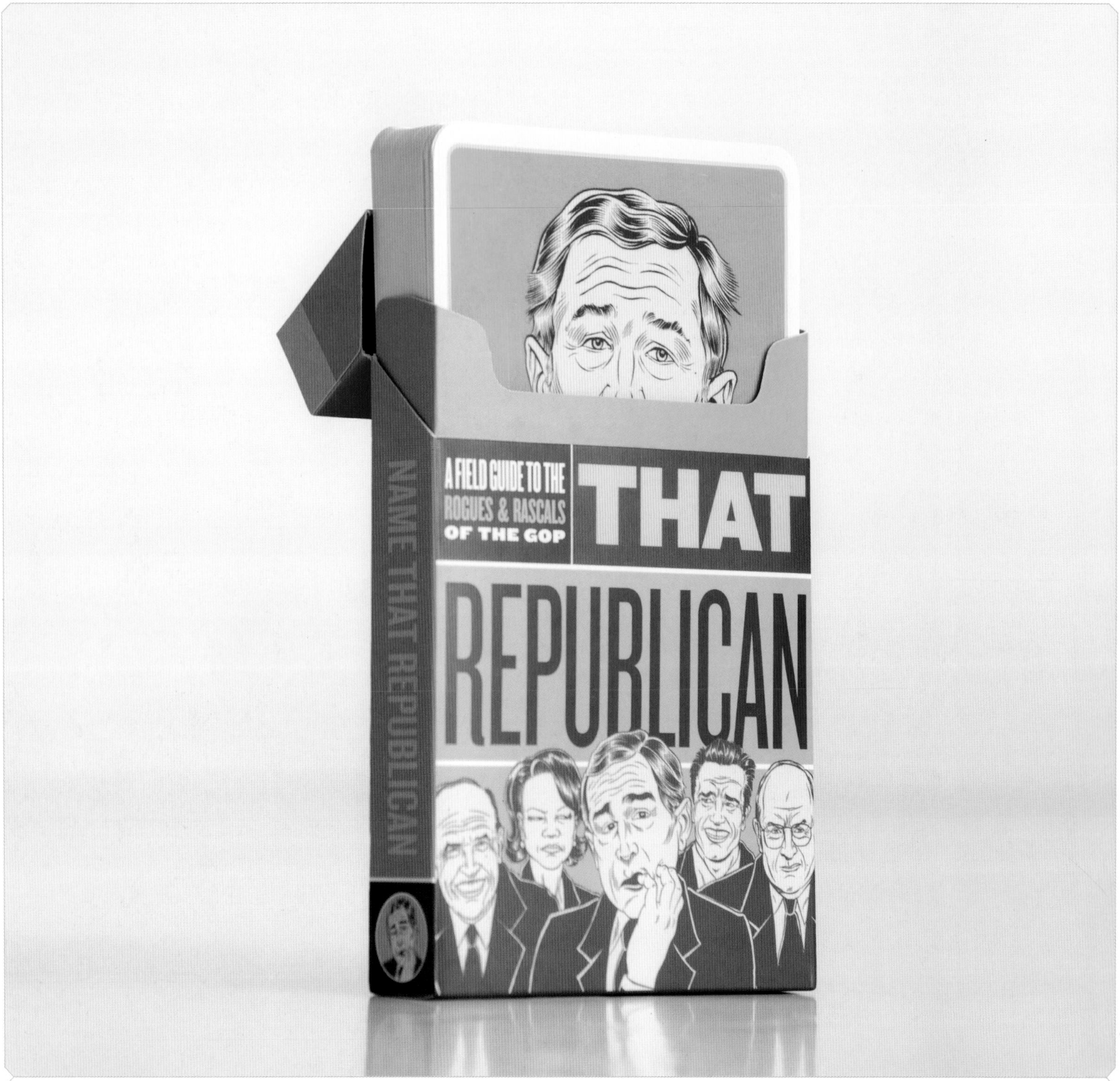
A FIELD GUIDE TO THE
ROGUES & RASCALS
OF THE GOP
THAT
REPUBLICAN
NAME THAT REPUBLICAN

CHRONICLE BOOKS
USA

0001

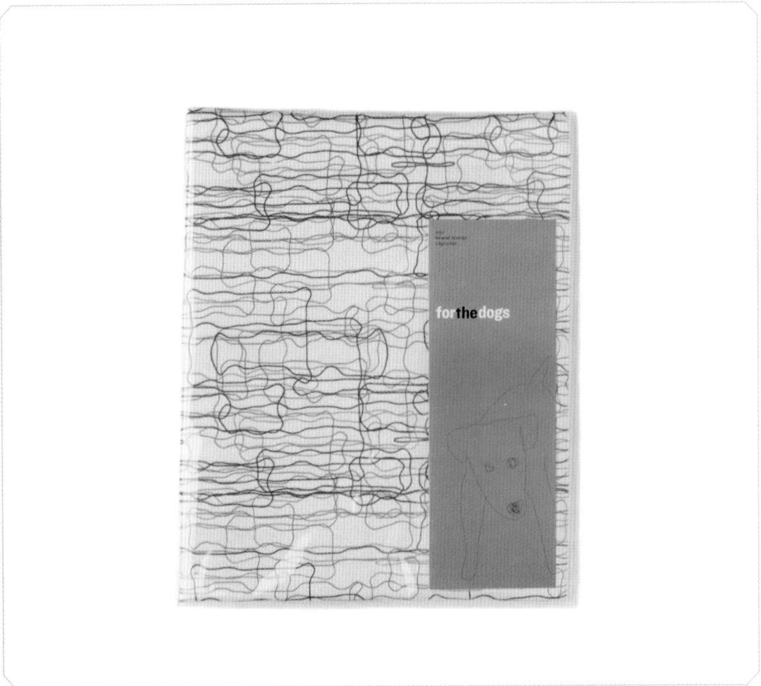

CONCRETE DESIGN COMMUNICATIONS
CANADA

0002

CONCRETE DESIGN COMMUNICATIONS
CANADA

0003

CONCRETE DESIGN COMMUNICATIONS
CANADA

0004

PRINCIPLE
USA

0005

GRIP
USA

0006

GRIP
USA

0007

GRIP
USA

0008

GRIP
USA

0009

TEAFOAM
a foaming cleanser
{with green tea}
6.7 FLUID OUNCES {200ML}
DUNDEEDERMATOLOGY
TEAWASH
a gentle cleanser
{fortified with green tea}
6.7 FLUID OUNCES {200ML}
DUNDEEDERMATOLOGY

GRIP
USA

0010

Alto Moncayo
2004 Garnacha
CAMPO DE BORJA - Denominación de Origen
BULBUENTE - ESPAÑA
Alto Moncayo
750 ML x 6 BOTTLES PRODUCT OF SPAIN
Alto Moncayo
2002 Garnacha
Alto Moncayo
2004 Garnacha

GO BIG
USA

0011

FIFTY STRATEGY & CREATIVE
CANADA
0012

FIFTY STRATEGY & CREATIVE
CANADA
0013

FIFTY STRATEGY & CREATIVE
CANADA
0014

FIFTY STRATEGY & CREATIVE
CANADA
0015

FIFTY STRATEGY & CREATIVE
CANADA
0016

FIFTY STRATEGY & CREATIVE
CANADA
0017

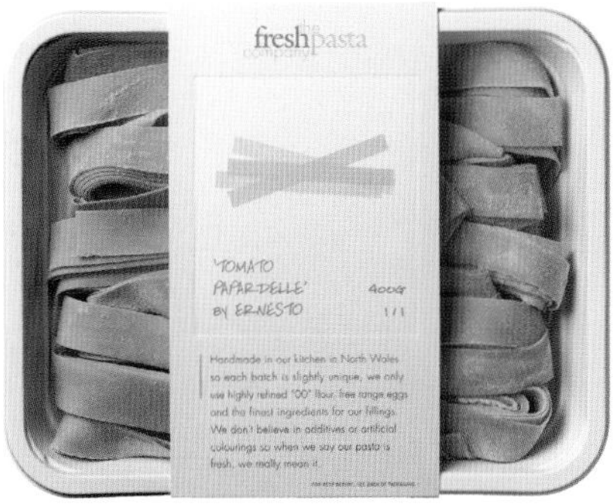

P&W
USA
0018

P&W
USA
0019

FIFTY STRATEGY & CREATIVE
CANADA
0020

CONCRETE DESIGN COMMUNICATIONS
CANADA
0021

CONCRETE DESIGN COMMUNICATIONS
CANADA
0022

CONCRETE DESIGN COMMUNICATIONS
CANADA
0023

CONCRETE DESIGN COMMUNICATIONS
CANADA
0024

CONCRETE DESIGN COMMUNICATIONS
CANADA
0025

CONCRETE DESIGN COMMUNICATIONS
CANADA
0026

CONCRETE DESIGN COMMUNICATIONS
CANADA
0027

DESIGN AHEAD
GERMANY
0028

DESIGN AHEAD
GERMANY
0029

CONCRETE DESIGN COMMUNICATIONS
CANADA

0030

FLIGHT 001
F-001
NOMAD SHOPPER FOR TRAVELERS DO NOT DISCARD

FLIGHT 001
USA

0031

FLIGHT PAK
FLIGHT 001

FLIGHT 001
USA

0032

FLIGHT 001
USA

0033

FIVE STAR HOTEL
TRAVELING ON A BUDGET? UPGRADE WITH FLIGHT 001'S EXCLUSIVE FIVE STAR HOTEL KIT. INDULGE YOURSELF WITH ALL THE AMENITIES OF A LUXURY SUITE INCLUDING A DO NOT DISTURB SIGN AND A COMPLIMENTARY CHOCOLATE FOR YOUR PILLOW.
CINQ ESTRELLAS
CINQUE
FIVE STAR

FLIGHT 001
USA

0034

FLIGHT 001
USA

0035

212-BIG-BOLT
USA

0036

212-BIG-BOLT
USA

0037

212-BIG-BOLT
USA

0038

212-BIG-BOLT
USA

0039

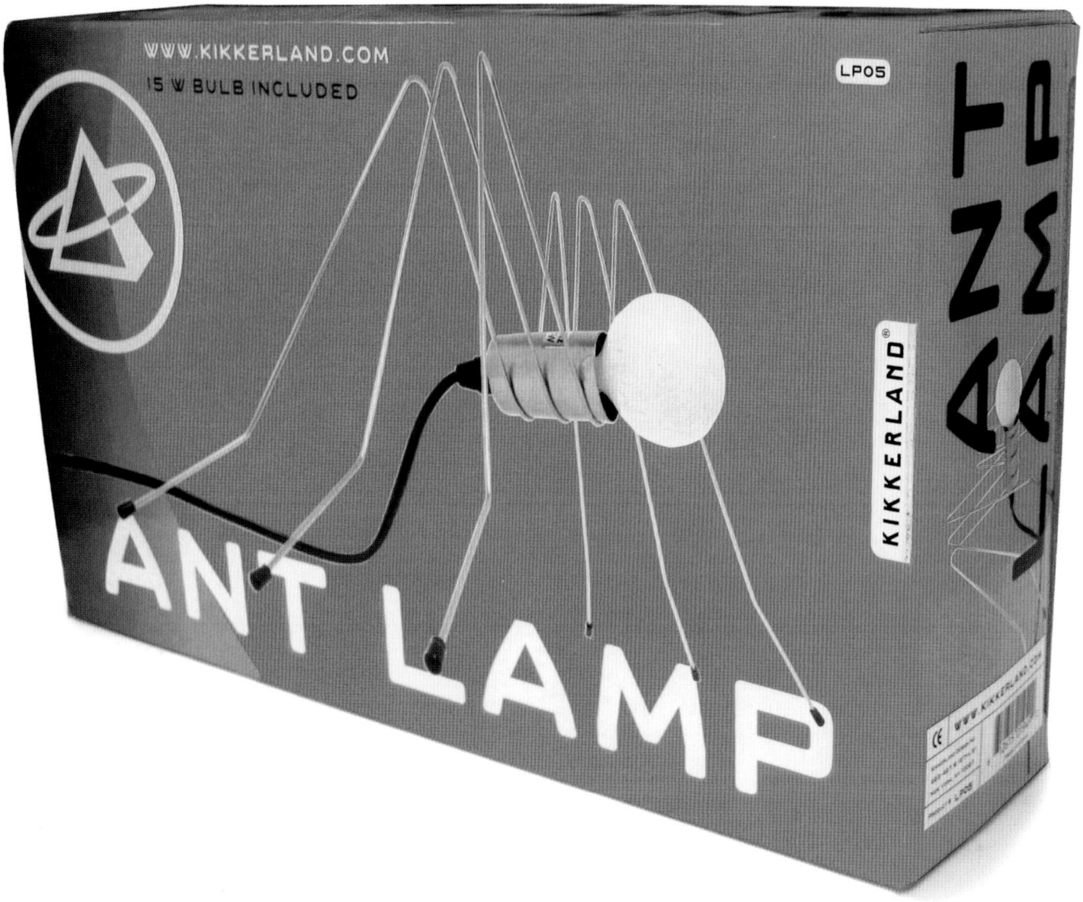
WWW.KIKKERLAND.COM
15 W BULB INCLUDED
LP05
ANT LAMP
KIKKERLAND®
ANT LAMP

212-BIG-BOLT
USA

0040

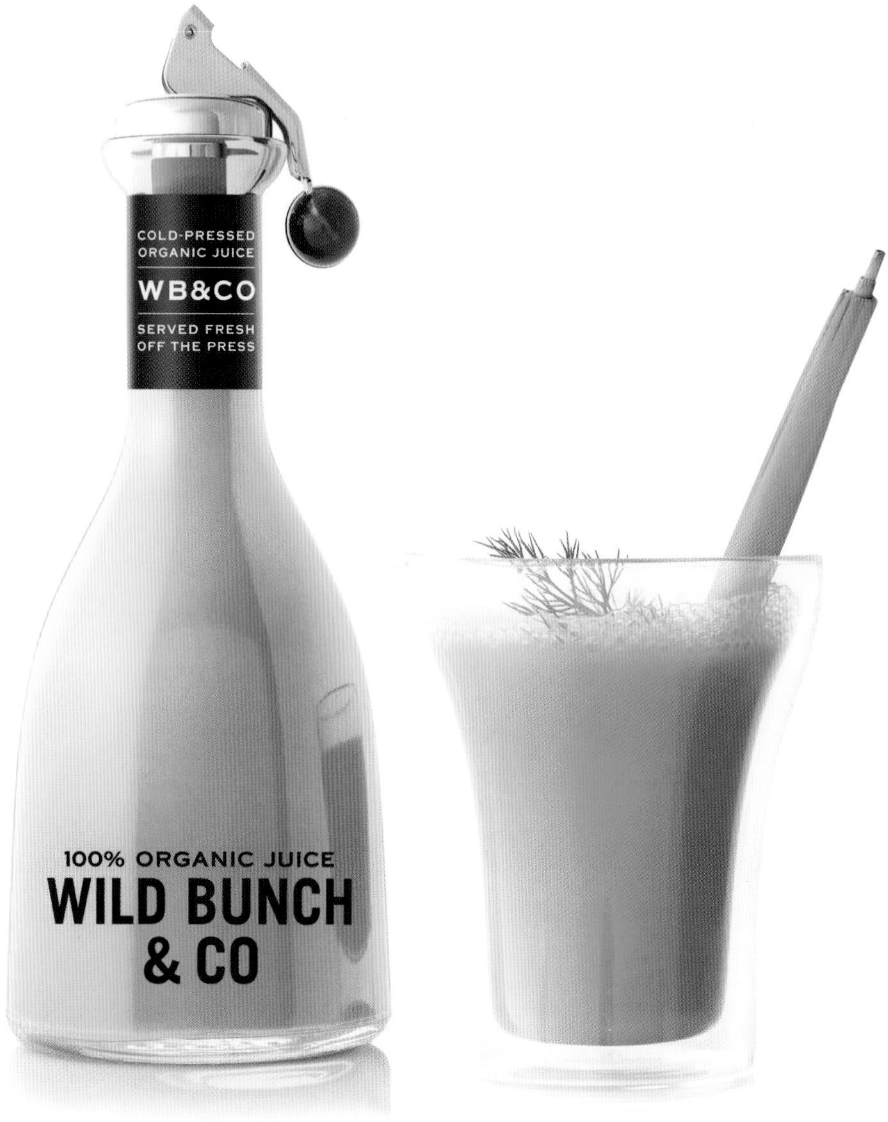
COLD-PRESSED
ORGANIC JUICE
WB&CO
SERVED FRESH
OFF THE PRESS
100% ORGANIC JUICE
WILD BUNCH
& CO

SEED
SINGAPORE

0041

BEET IT
WILD BUNCH
& CO

SEED
SINGAPORE
0042

EASY
PEAZY
WILD BUNCH
& CO

SEED
SINGAPORE
0043

IRON
MAIDEN
WILD BUNCH
& CO

SEED
SINGAPORE
0044

SAVAGE
CABBAGE
WILD BUNCH
& CO

SEED
SINGAPORE
0045

ROOTINE
CHECK
WILD BUNCH
& CO

SEED
SINGAPORE
0046

PURE
CARROT
WILD BUNCH
& CO

SEED
SINGAPORE
0047

Waitrose
Soft Malatya Apricots
Dried pitted apricots
Selected for size and quality, these naturally sweet apricots are grown in the favourable climate of the Malatya region of Turkey.

TURNER DUCKWORTH
USA
0048

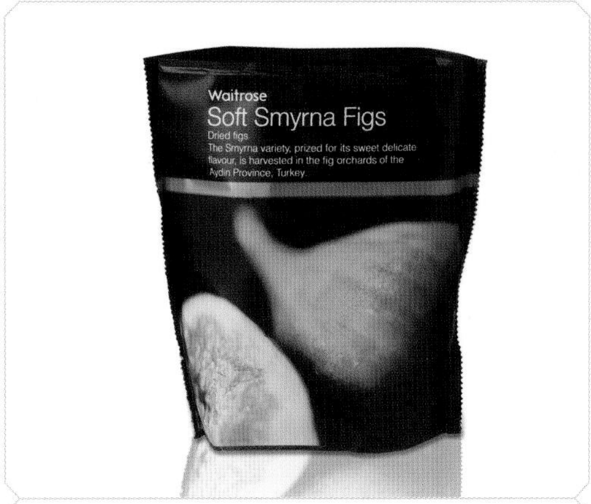
Waitrose
Soft Smyrna Figs
Dried figs
The Smyrna variety, prized for its sweet delicate flavour, is harvested in the fig orchards of the Aydin Province, Turkey.

TURNER DUCKWORTH
USA
0049

Waitrose
Soft d'Agen Prunes
Pitted prunes
Growing conditions of the orchards around Agen, South West France, provide a perfect climate for producing a unique rich flavour.

TURNER DUCKWORTH
USA
0050

OUTSET, INC.
USA
0051

OUTSET, INC.
USA
0052

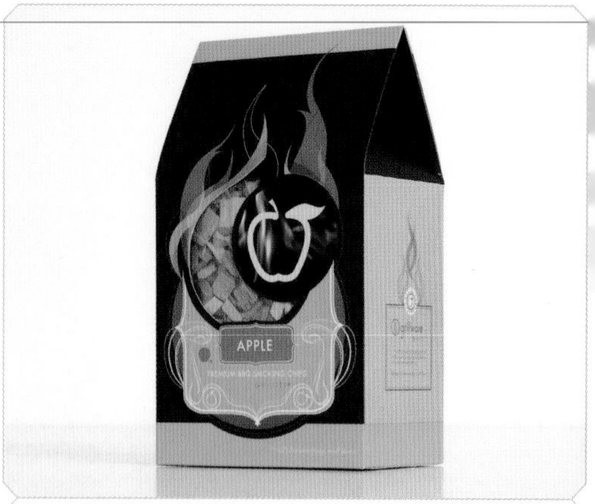

OUTSET, INC.
USA
0053

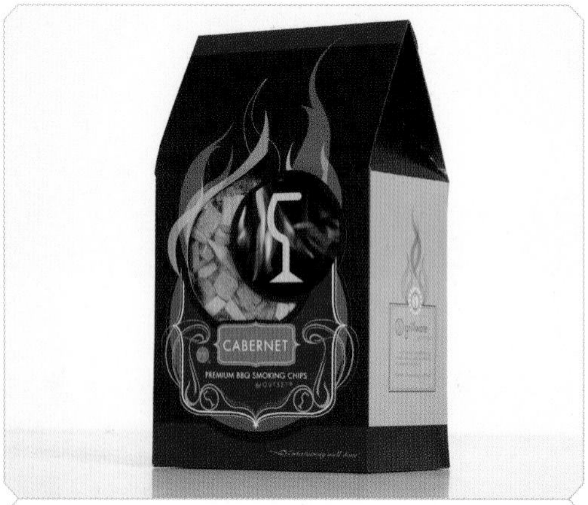

OUTSET, INC.
USA
0054

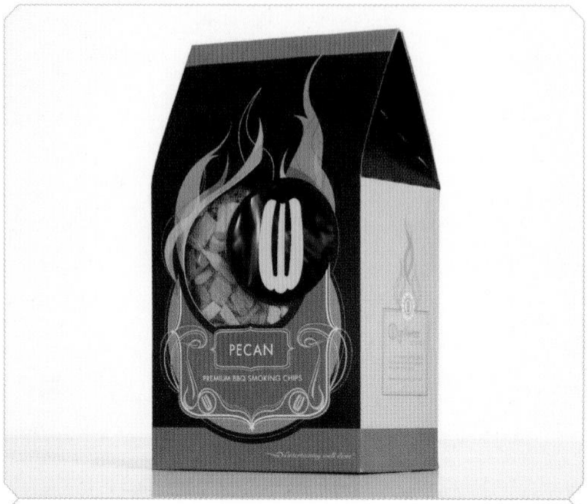

OUTSET, INC.
USA
0055

OUTSET, INC.
USA
0056

OUTSET, INC.
USA
0057

OUTSET, INC.
USA
0058

IRVING
UK
0059

ICE BUCKET
& TONGS
Entertaining well done™

OUTSET, INC.
USA

0060

VARIETAL: SAUVIGNON BLANC
ROOTSTOCK: ORIGINAL UNGRAFTED
Root:1
Chile is a true rarity in the wine world. Unique geographic & climatic forces have allowed it to remain one of very few grape growing regions in the world where the original European rootstocks survive, unaffected by phylloxera – the disease that forced grape growers worldwide to graft vines onto generic rootstocks. Chile's isolation, protected by the mighty Andes to the East and the vast expanse of the Pacific Ocean to the West, ensures that grape vines can remain on the original rootstock, in the purest form. These same geographic characteristics provide Chile's fertile central region with optimal climate & soil conditions to produce consistent and outstanding grapes each year. This Root:1 Sauvignon Blanc is crafted exclusively from grapes grown on the original, ungrafted root systems tended by our Master Winemakers. This gives the wine its pure, rich fruit flavors & aromas.
Root:1™
[THE ORIGINAL UNGRAFTED]
SAUVIGNON BLANC
CASABLANCA VALLEY, CHILE 2007
VARIETAL: CABERNET SAUVIGNON
ROOTSTOCK: ORIGINAL UNGRAFTED
Root:1
Root:1™
[THE ORIGINAL UNGRAFTED]
CABERNET SAUVIGNON
COLCHAGUA VALLEY, CHILE 2006

TURNER DUCKWORTH
USA

0061

ARLO
USA

0062

DESIGN BRIDGE
UK

0063

KYM ABRAMS DESIGN
USA

0064

DESIGN BRIDGE
UK

0065

TABLE BRIWE INK
USA

0066

MICHAEL OSBORNE DESIGN
USA

0067

TURNSTYLE
USA

0068

MICHAEL OSBORNE DESIGN
USA

0069

BOOTLEG
MADE IN ITALY
BOOTLEG
SOUTHERN RED
MADE IN ITALY
BOOTLEG
CHIANTI
MADE IN ITALY
BOOTLEG
SAUVIGNON BLANC
MADE IN ITALY
BOOTLEG
PINOT GRIGIO
MADE IN ITALY

TURNER DUCKWORTH
USA

0070

saké2me
GREEN TEA
saké2me
–LIGHTLY– SPARKLING
ASIAN PEAR
MADE FROM IMPORTED
JAPANESE
JUNMAI SAKÉ
SAKE WITH NATURAL FLAVOR
saké2me
–LIGHTLY– SPARKLING
GINGER MANGO
MADE FROM IMPORTED
JAPANESE
JUNMAI SAKÉ
WITH NATURAL FLAVORS AND BETA CAROTENE AND CARROT JUICE
saké2me
YUZU CITRUS
MADE FROM IMPORTED
JAPANESE
JUNMAI SAKÉ

MICHAEL OSBORNE DESIGN
USA

0071

THE COLLECTIVE DESIGN CONSULTANTS
AUSTRALIA
0072

THE COLLECTIVE DESIGN CONSULTANTS
AUSTRALIA
0073

STUDIOBENBEN
USA
0074

THE COLLECTIVE DESIGN CONSULTANTS
AUSTRALIA
0075

TURNER DUCKWORTH
USA
0076

TURNER DUCKWORTH
USA
0077

TURNER DUCKWORTH
USA
0078

TURNER DUCKWORTH
USA
0079

TURNER DUCKWORTH
USA
0080

TURNER DUCKWORTH
USA
0081

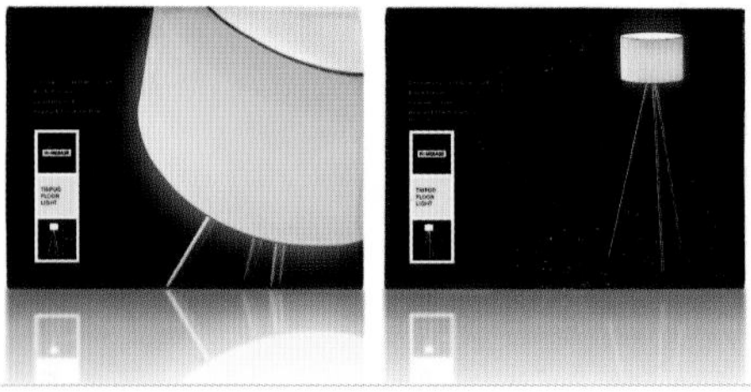

TURNER DUCKWORTH
USA
0082

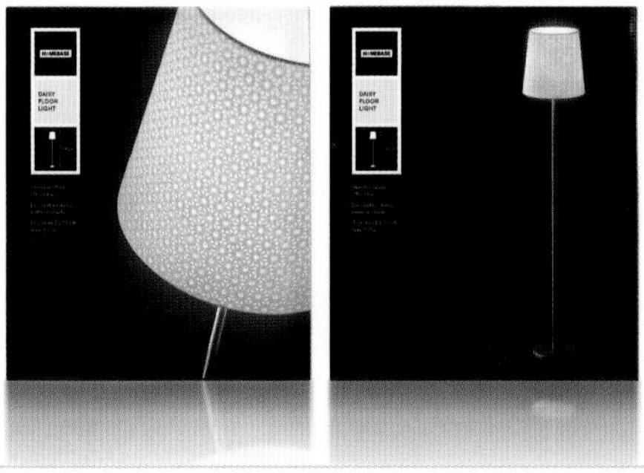

TURNER DUCKWORTH
USA
0083

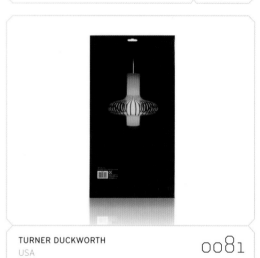

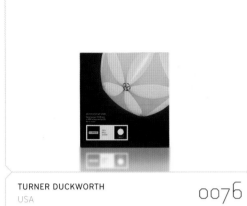

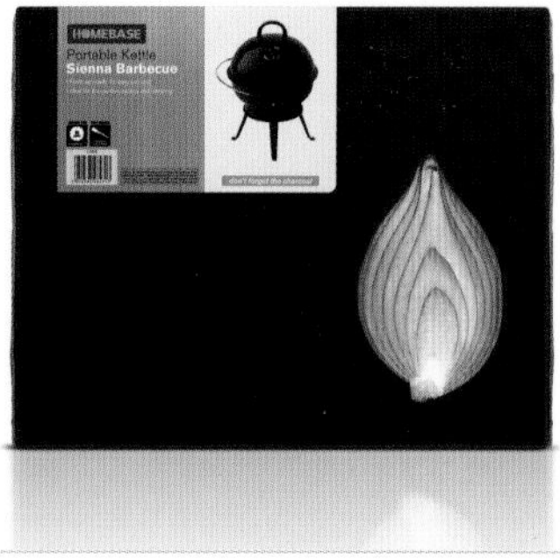

TURNER DUCKWORTH
USA
0084

HOMEBASE
Deluxe Trolley
Sicilia Gas Barbecue
• 8.8 kW power with 3 burners and 2.9 kW side burner
• Porcelain coated grill and warming rack for easy cleaning
• 27mm clip on propane regulator
don't forget the calor gas

TURNER DUCKWORTH
USA

0085

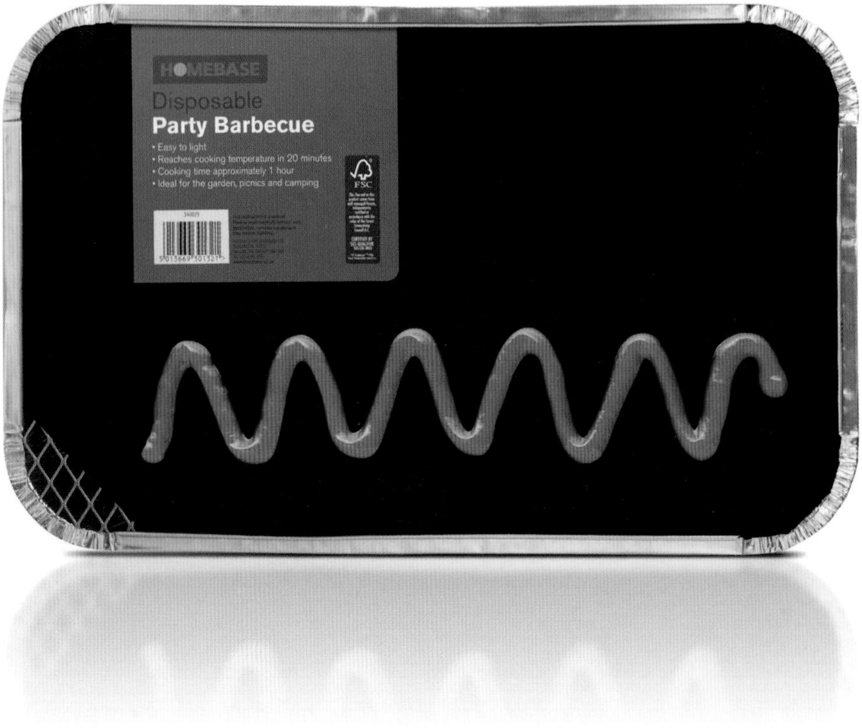
HOMEBASE
Disposable
Party Barbecue
• Easy to light
• Reaches cooking temperature in 20 minutes
• Cooking time approximately 1 hour
• Ideal for the garden, picnics and camping
FSC

TURNER DUCKWORTH
USA

0086

TURNER DUCKWORTH
USA

0087

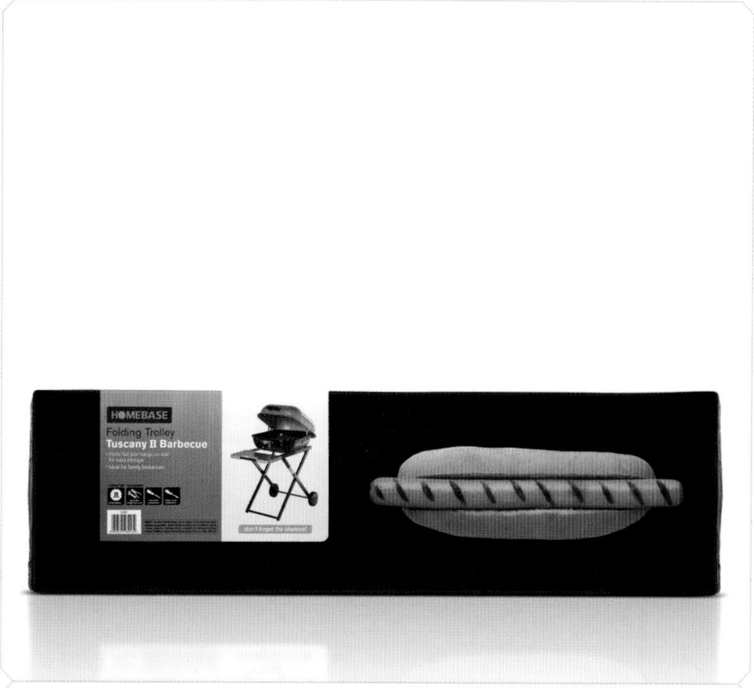

TURNER DUCKWORTH
USA

0088

TURNER DUCKWORTH
USA

0089

TURNER DUCKWORTH
USA

0090

TURNER DUCKWORTH
USA
0091

TURNER DUCKWORTH
USA
0092

TURNER DUCKWORTH
USA
0093

TURNER DUCKWORTH
USA
0094

MUCCA DESIGN CORP.
USA

0095

Waitrose
SPAGHETTI HOOPS
IN TOMATO SAUCE
Waitrose
LONG SPAGHETTI
IN TOMATO SAUCE
Waitrose
BAKED BEANS
IN TOMATO SAUCE

TURNER DUCKWORTH
USA

0096

TURNER DUCKWORTH
USA
0097

TURNER DUCKWORTH
USA
0098

TURNER DUCKWORTH
USA
0099

TURNER DUCKWORTH
USA
0100

TURNER DUCKWORTH
USA
0101

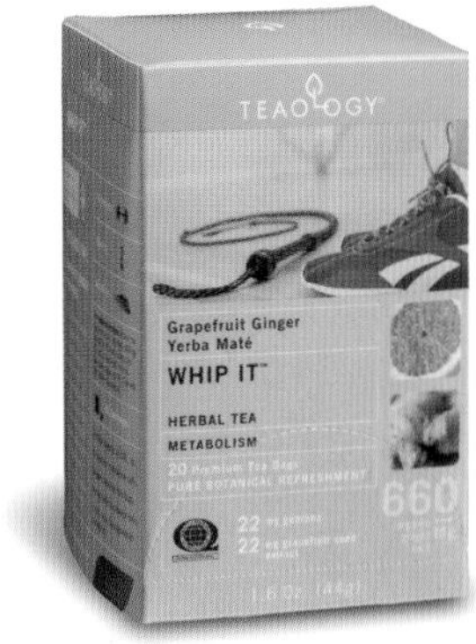

MARY HUTCHISON DESIGN LLC
USA
0102

P&W
UK
0103

P&W
UK
0104

P&W
UK
0105

SONSOLES LLORENS
SPAIN
0106

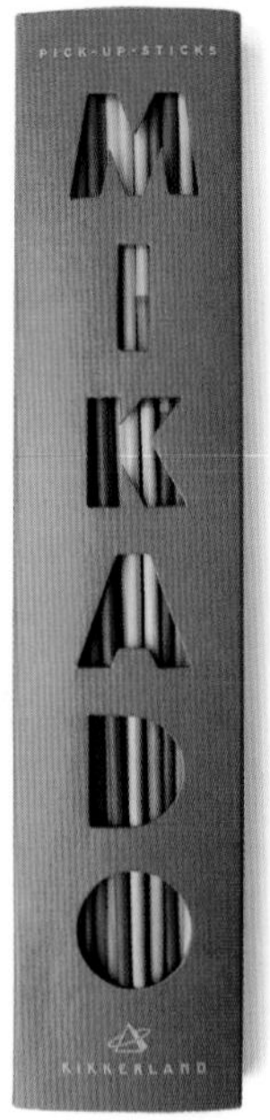

212-BIG-BOLT
USA
0107

212-BIG-BOLT
USA
0108

SHINE ADVERTISING
USA
0109

AMERICAN APPAREL
USA

0110

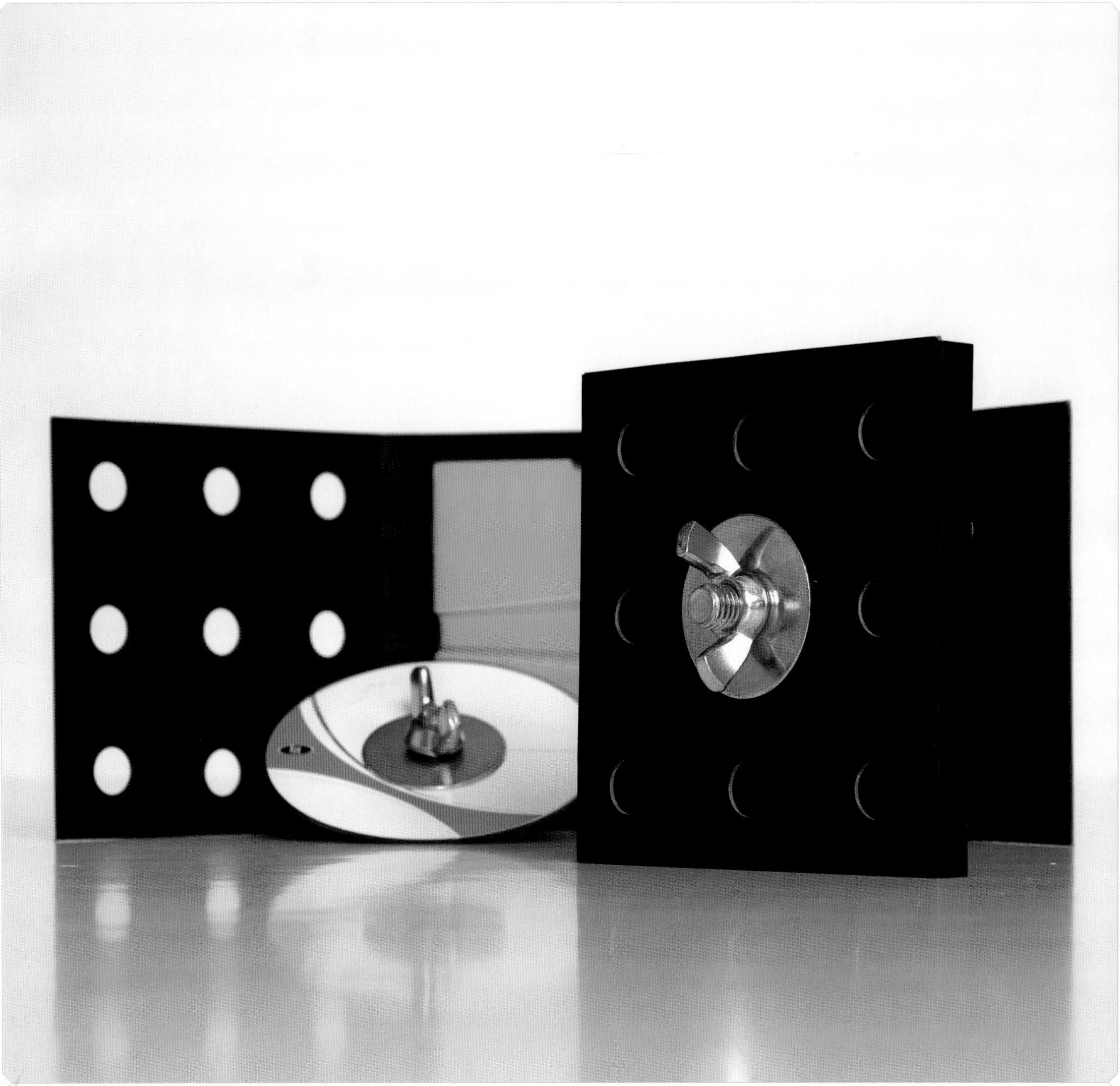

BFG
USA

0111

ZION GRAPHICS
SWEDEN
0112

ZION GRAPHICS
SWEDEN
0113

PINK BLUE BLACK & ORANGE CO., LTD.
THAILAND
0114

ZION GRAPHICS
SWEDEN
0115

ZION GRAPHICS
SWEDEN
0116

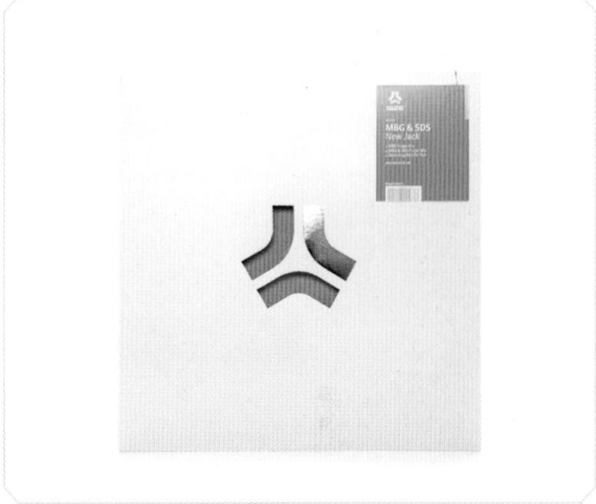

ZION GRAPHICS
SWEDEN
0117

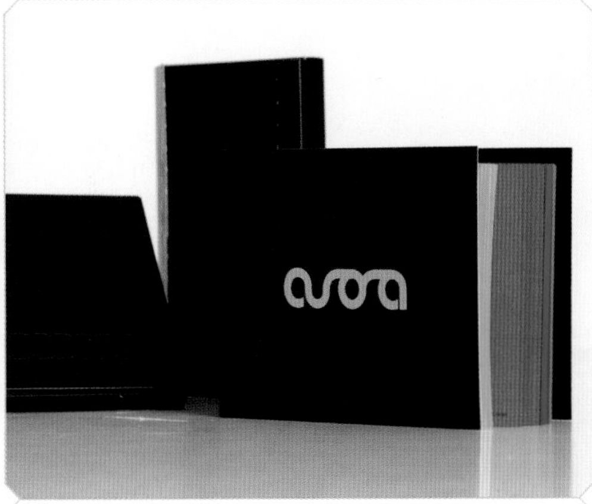

AURORA DESIGN
UK
0118

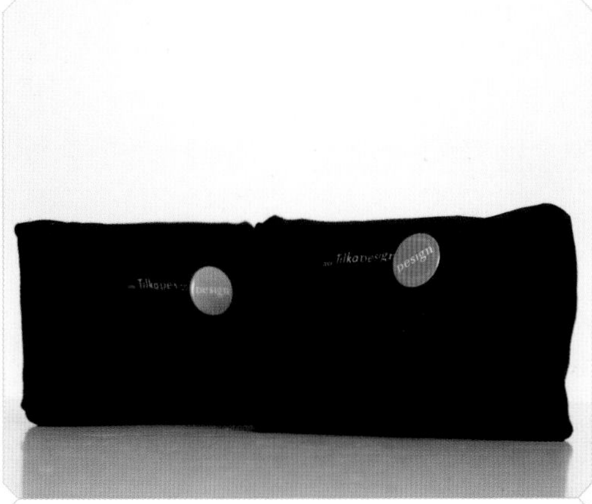

TILKA DESIGN
USA
0119

DECKER DESIGN
USA
0120

CONCRETE DESIGN COMMUNICATIONS
CANADA
0121

CONCRETE DESIGN COMMUNICATIONS
CANADA
0122

CONCRETE DESIGN COMMUNICATIONS
CANADA
0123

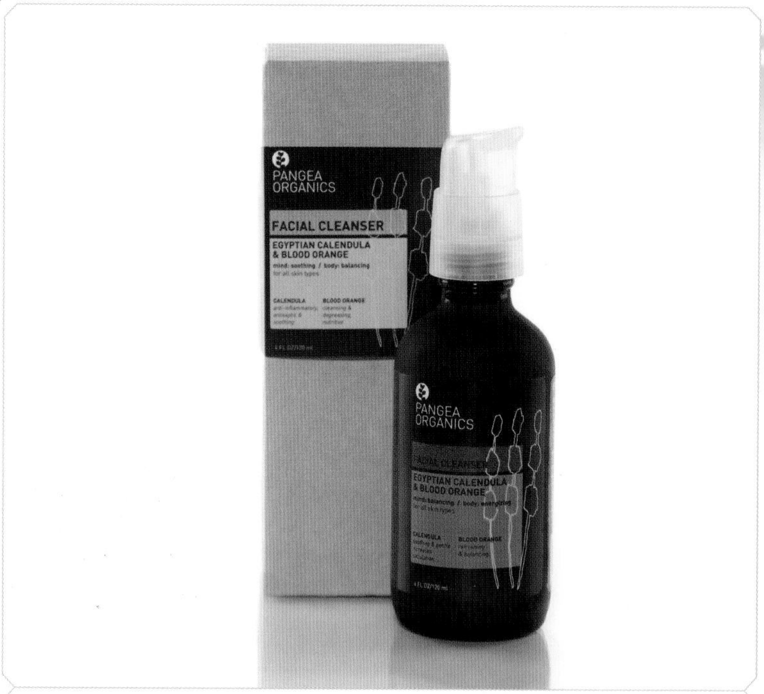

PANGEA ORGANICS
USA
0124

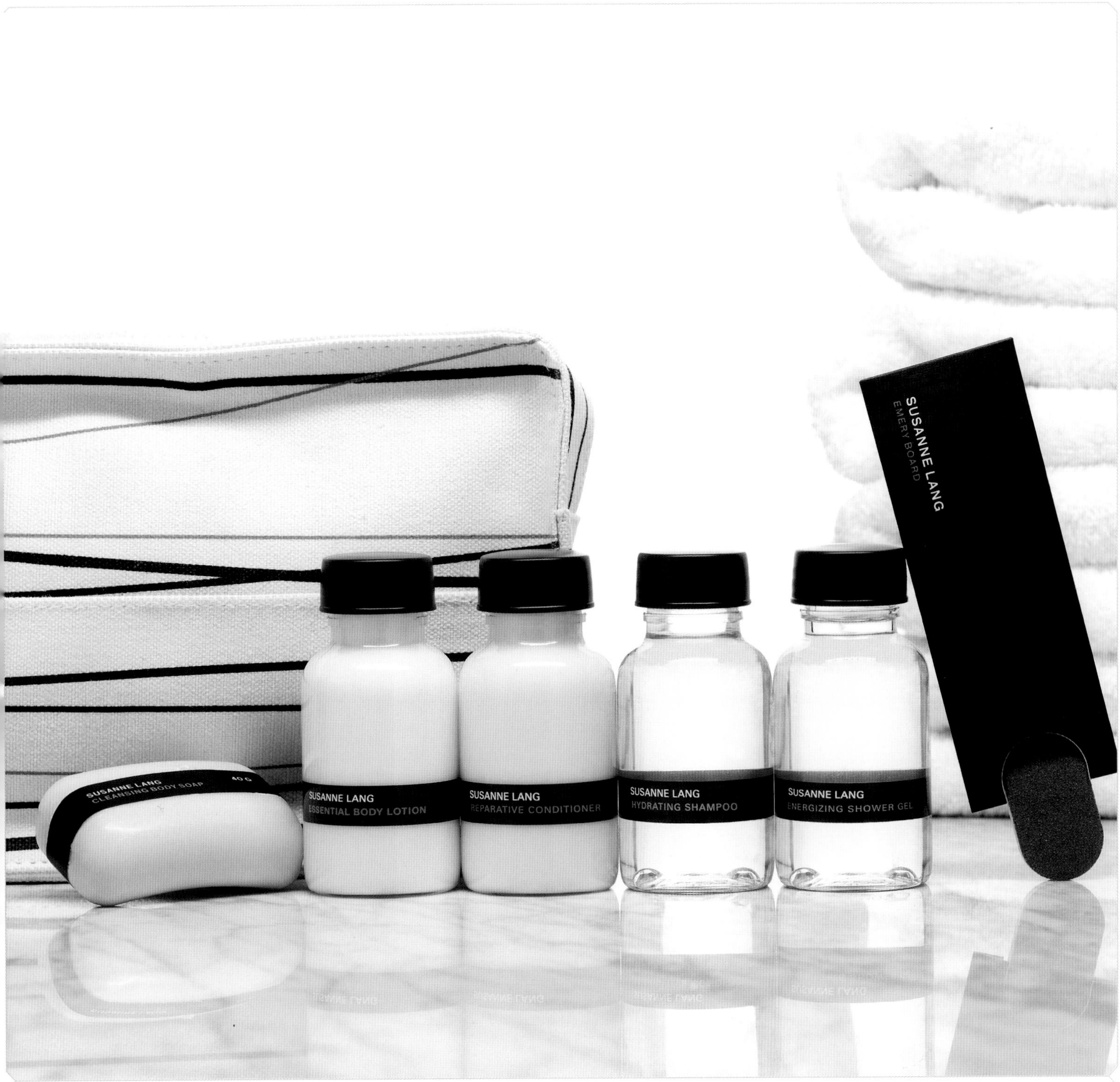
SUSANNE LANG
CLEANSING BODY SOAP
40 G
SUSANNE LANG
ESSENTIAL BODY LOTION
SUSANNE LANG
REPARATIVE CONDITIONER
SUSANNE LANG
HYDRATING SHAMPOO
SUSANNE LANG
ENERGIZING SHOWER GEL
SUSANNE LANG
EMERY BOARD

CONCRETE DESIGN COMMUNICATIONS
CANADA

0125

PANGEA
ORGANICS
MASSAGE & BODY OIL
MALAGASY GINGER
WITH LEMONGRASS
mind: invigorating / body: rejuvenating
LEMONGRASS
ROSEMARY
GINGER
4 FL OZ/120 ml

PANGEA
ORGANICS
MASSAGE & BODY OIL
PYRENEES LAVENDER
WITH CARDAMOM
mind: clarifying / body: warming
rejuvenating oil
LAVENDER
CARDAMOM
4 FL OZ/120 ml

PANGEA
ORGANICS
MASSAGE & BODY OIL
ITALIAN WHITE SAGE
& GERANIUM
mind: grounding / body: balancing
restorative oil
WHITE SAGE
GERANIUM
4 FL OZ/120 ml

PANGEA ORGANICS
USA

0126

PANGEA ORGANICS
USA
0127

PANGEA ORGANICS
USA
0128

PANGEA ORGANICS
USA
0129

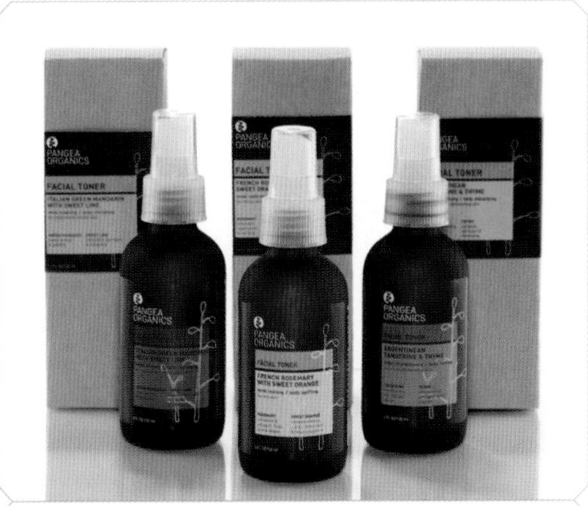

PANGEA ORGANICS
USA
0130

PANGEA ORGANICS
USA
0131

PANGEA ORGANICS
USA
0132

PANGEA ORGANICS
USA
0133

PANGEA ORGANICS
USA
0134

PANGEA ORGANICS
USA
0135

PANGEA ORGANICS
USA
0136

PANGEA ORGANICS
USA
0137

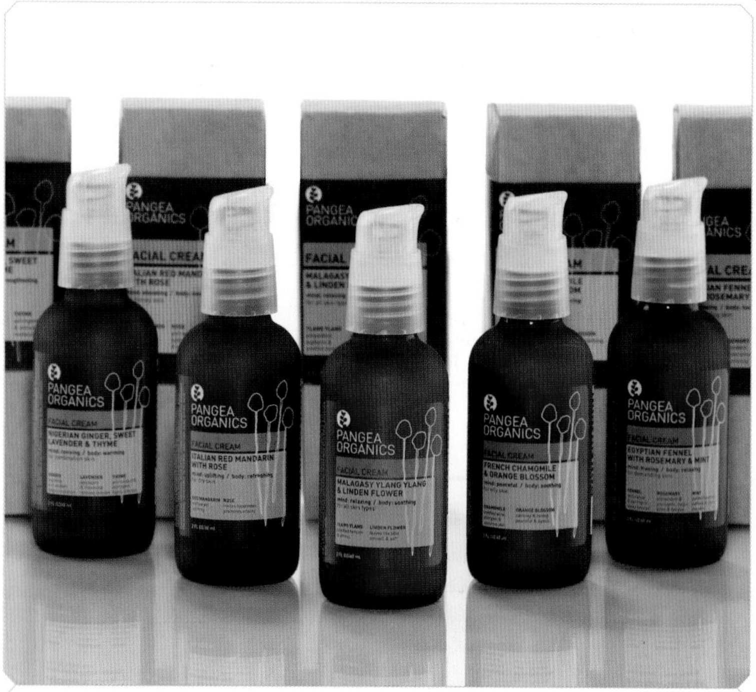

PANGEA ORGANICS
USA
0138

PANGEA ORGANICS
USA
0139

PANGEA ORGANICS
ECOCENTRIC
LIPCARE
LIPCARE
EGYPTIAN FENNEL WITH
GRAPEFRUIT & SWEET ORANGE
GRAPEFRUIT
antioxidant
euphoric
SWEET ORANGE
soothing
vitamin-rich
PYRENEES LAVENDER
WITH CARDAMOM
LAVENDER
healing
calming
CARDAMOM
antioxidant
reviving
ITALIAN RED MANDARIN
WITH ROSE
RED MANDARIN
soothing
vitamin-rich
ROSE
hydrating
protecting

PANGEA ORGANICS
USA

0140

eucalyptus mint
menthe eucalyptus
NON-TOXIC CLEAN
planet friendly gently abrasive cleaner
LE SCRUB™
récurant éco non toxique pour salle de bain
method
eucalyptus mint | menthe eucalyptus
NON-TOXIC CLEAN
planet friendly toilet bowl cleaner
LIL' BOWL BLU™
nettoyant cuvettes non-toxique et éco
method

METHOD
USA

0141

METHOD
USA
0142

METHOD
USA
0143

MYINT DESIGN
USA
0144

METHOD
USA
0145

METHOD
USA
0146

METHOD
USA
0147

METHOD
USA
0148

JOED DESIGN INC.
USA
0149

PEARLFISHER
UK
0150

VINTAGE

LMS DESIGN INC.
USA
0151

PIZZA NOVA
PIZZA NOVA
PIZZA NOVA
PIZZA NOVA
PIZZA NOVA
PIZZA NOVA

CONCRETE DESIGN COMMUNICATIONS
CANADA
0152

PIZZA NOVA

CONCRETE DESIGN COMMUNICATIONS
CANADA
0153

TESCO
INGREDIENTS
SEA SALT FLAKES 200g℮
Roasted Sea Bass: Take a whole sea bass, slash twice on each side and insert fresh rosemary. Rub fish with olive oil and sprinkle with Sea Salt. Roast in the oven.

P&W
UK
0154

TESCO
INGREDIENTS
KAFFIR LIME LEAVES 2g

P&W
UK
0155

TESCO
INGREDIENTS
VANILLA PODS
2 PODS

P&W
UK
0156

TESCO
INGREDIENTS
TOASTED PINE NUTS
Boardroom Chicken: Fry Pine Nuts in a dry pan, add chicken pieces and flash fry. Place in an ovenproof dish with olive oil and sundried tomatoes. Bake in the oven.

P&W
UK
0157

TESCO
INGREDIENTS
DRIED PORCINI MUSHROOMS 40g
Pasta Porcini: Soak Dried Porcini Mushrooms in hot water and drain. Sauté garlic in olive oil, add salt and pepper, tomatoes and the mushrooms. Simmer. Serve with cooked pasta.

P&W
UK
0158

TESCO
INGREDIENTS
WHITE BALSAMIC CONDIMENT
250ml℮

P&W
UK
0159

CONCRETE DESIGN COMMUNICATIONS
CANADA

0160

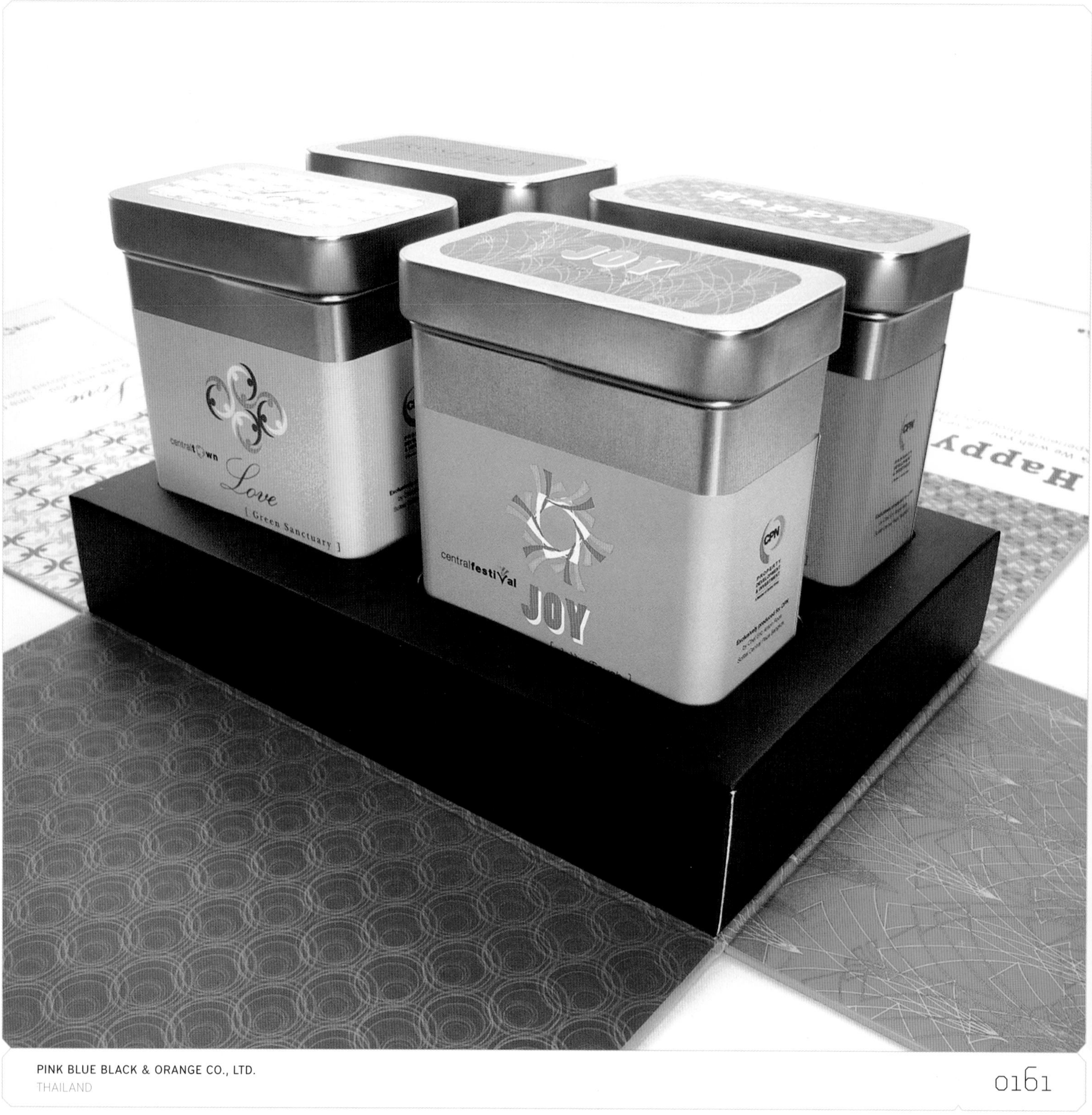
Happy
JOY
centraltown
Love
[Green Sanctuary]
centralfestival
JOY
CPN

PINK BLUE BLACK & ORANGE CO., LTD.
THAILAND

0161

PINK BLUE BLACK & ORANGE CO., LTD.
THAILAND
0162

PINK BLUE BLACK & ORANGE CO., LTD.
THAILAND
0163

SALVARTES DISEÑO Y PUBLICIDAD
SPAIN
0164

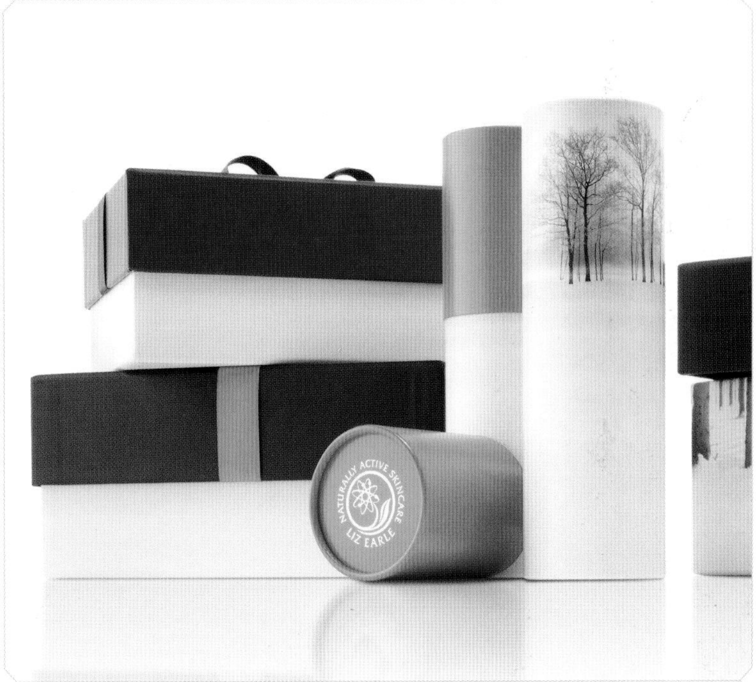

TURNER DUCKWORTH
USA
0165

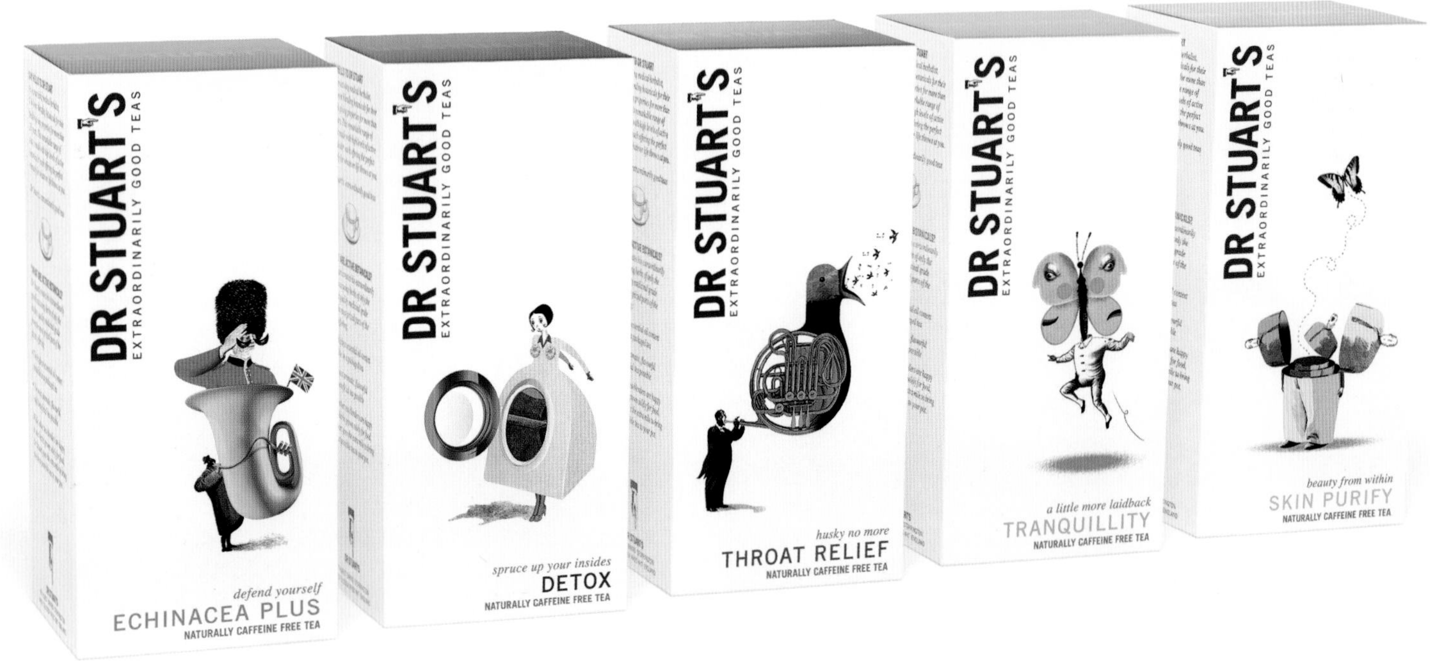
DR STUART'S
EXTRAORDINARILY GOOD TEAS
defend yourself
ECHINACEA PLUS
NATURALLY CAFFEINE FREE TEA
spruce up your insides
DETOX
NATURALLY CAFFEINE FREE TEA
husky no more
THROAT RELIEF
NATURALLY CAFFEINE FREE TEA
a little more laidback
TRANQUILLITY
NATURALLY CAFFEINE FREE TEA
beauty from within
SKIN PURIFY
NATURALLY CAFFEINE FREE TEA

PEARLFISHER
UK

0166

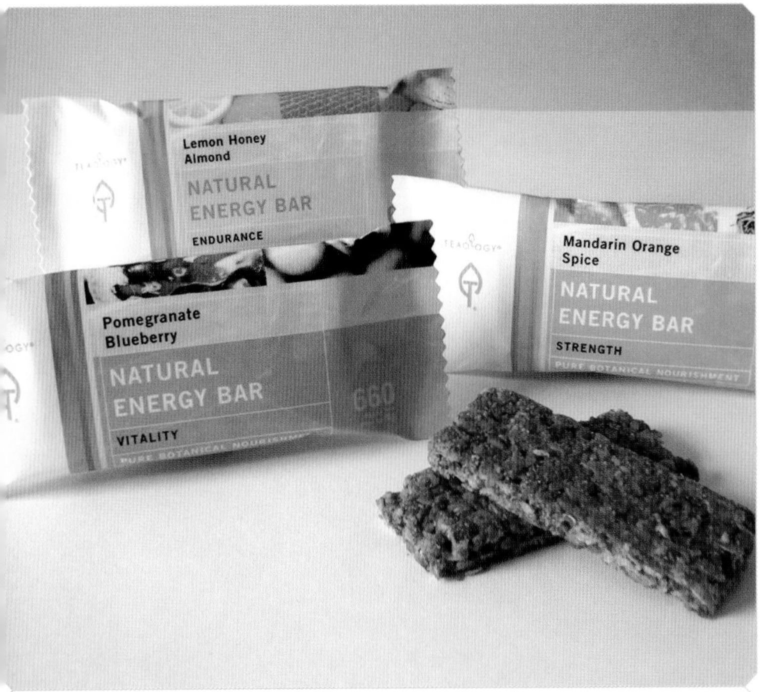

MARY HUTCHISON DESIGN LLC
USA
0167

MORLA DESIGN
USA
0168

CONCRETE DESIGN COMMUNICATIONS
CANADA
0169

CONCRETE DESIGN COMMUNICATIONS
CANADA
0170

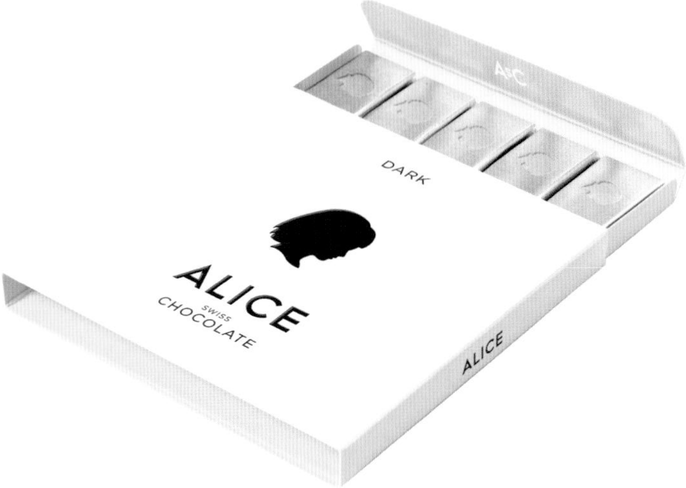

ALICE CHOCOLATE LLC
USA
0171

TURNSTYLE
USA
0172

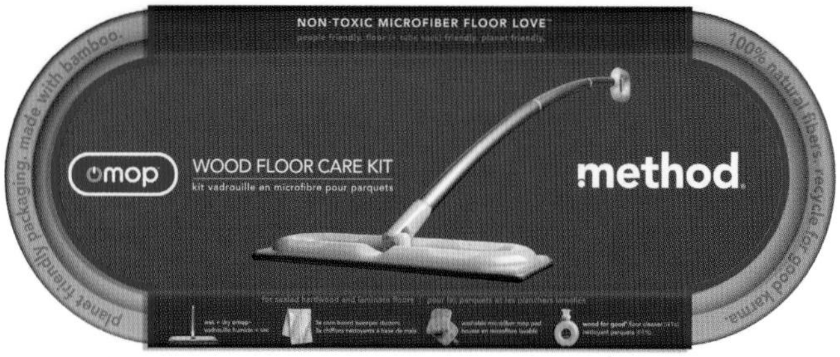

METHOD
USA
0173

HORNALL ANDERSON DESIGN
USA
0174

DEFINITIONS
Sterling
WAHL
POWERFUL ROTARY MOTOR PERFORMANCE
T-SHAPED BLADES ADJUST TO ZERO-OVERLAP
LIGHTWEIGHT – LESS THAN 5" IN LENGTH
THE ULTRA-CLOSE CUTTING TRIMMER
WAHL CLIPPER CORPORATION

GRIP
USA

BRAIS
TENDER BEEF SHOR
AND VEAL STOCK F
WALL
FOOD COM
HANDCRAFTED ON
LIME & CHILI INFUSED FRESH FRUIT
MANGO, PINEAPPLE, AND STAR FRUIT SUSPENDED IN
A LIME AND CHILI INFUSED SYRUP
WALLY'S
FOOD COMPANY
ISRAELI COUSCOUS & VEGETABLE CONFETTI
ISRAELI COUSCOUS TOSSED WITH ZUCCHINI, SQUASH
AND BELL PEPPER SERVED WITH BULGARIAN FETA
ALLY'S
COMPANY

PHILIPPE BECKER DESIGN
USA

0176

TURNER DUCKWORTH
USA

0177

TURNER DUCKWORTH
USA

0178

TURNER DUCKWORTH
USA

0179

TURNER DUCKWORTH
USA

0180

PINK BLUE BLACK & ORANGE CO., LTD.
THAILAND
0181

PINK BLUE BLACK & ORANGE CO., LTD.
THAILAND
0182

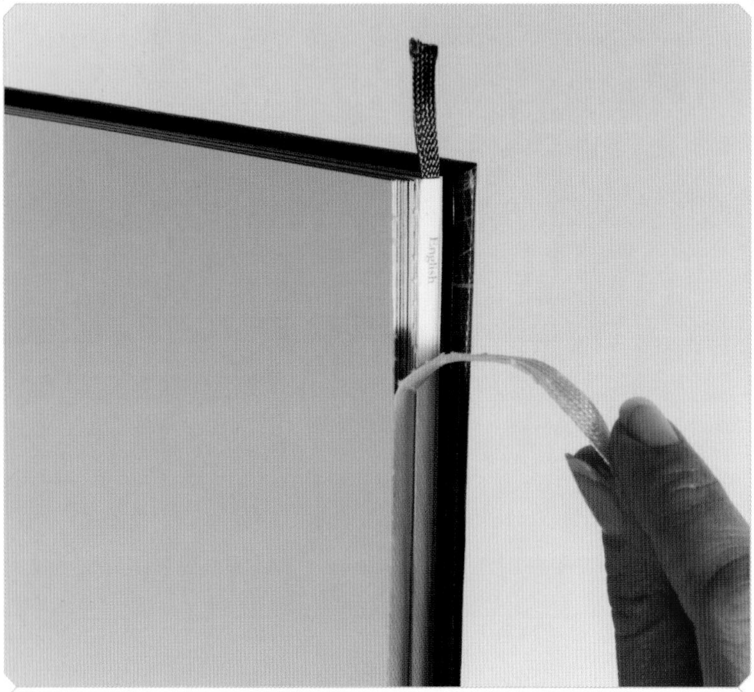

PAPA, INC.
CROATIA
0183

PAPA, INC.
CROATIA
184

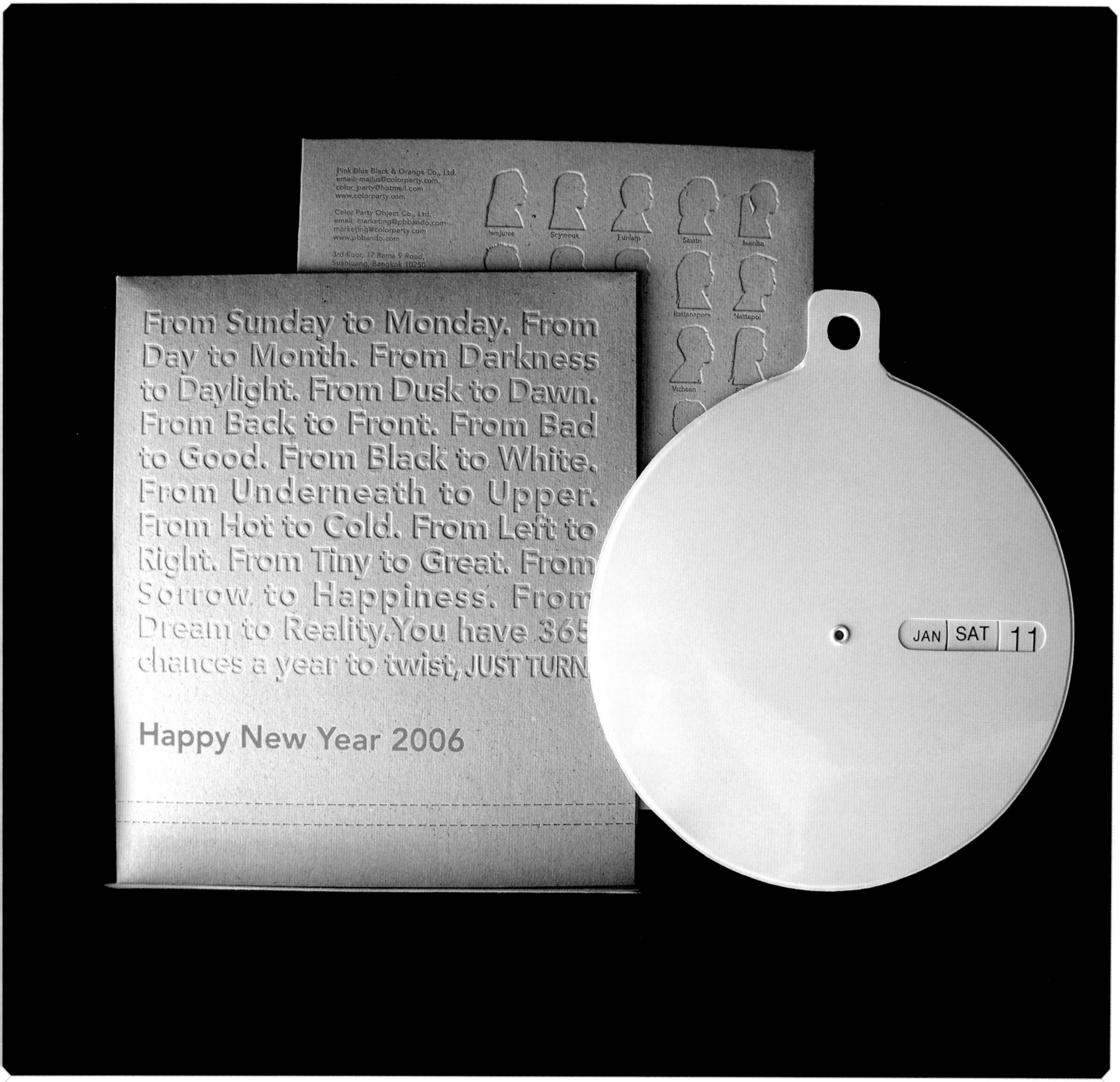
From Sunday to Monday. From Day to Month. From Darkness to Daylight. From Dusk to Dawn. From Back to Front. From Bad to Good. From Black to White. From Underneath to Upper. From Hot to Cold. From Left to Right. From Tiny to Great. From Sorrow to Happiness. From Dream to Reality.You have 365 chances a year to twist, JUST TURN.
Happy New Year 2006
JAN
SAT
11

PINK BLUE BLACK & ORANGE CO., LTD.
THAILAND

0185

gasp!

CUTE
ADORABLE
SNUGGLY
DELIGHTFUL
PRECIOUS
LOVABLE
DARLING
CLEVER
CHARMING
DISHY
IRRESISTIBLE
SWEET

funnel
PAPER
GOODS
FUNNEL
PAPER*
28 X 40
COMMERCIAL + FINE : ART ::::
Original printed patterns inspired by thoughts
of ancient, vinyl-padded, patio furniture,
flannel pjs and musty books. Created from
and intended to dress-up the events of our lives.
Good for blizzards, weddings, spring-flings,
anniversary toasts and lemonades by the

FUNNEL: ERIC KASS: UTILITARIAN + COMMERCIAL + FINE: ART
USA

0186

FRIENDSWITHYOU
USA

0187

FRIENDSWITHYOU
USA

0188

FRIENDSWITHYOU
USA

0189

FRIENDSWITHYOU
USA

0190

fruit from
the trees,
this
water
from a spring
mango, passion fruit & spring water

PEARLFISHER
UK
0191

this
water
quenches thirsts
for a living
blackcurrants & spring water

PEARLFISHER
UK
0192

this
water
is made from
fruit and clouds
lemon, lime & spring water

PEARLFISHER
UK
0193

there are 89
cranberries in
every bottle of
this
water
and 13 raspberries too
cranberries, raspberries & spring water

PEARLFISHER
UK
0194

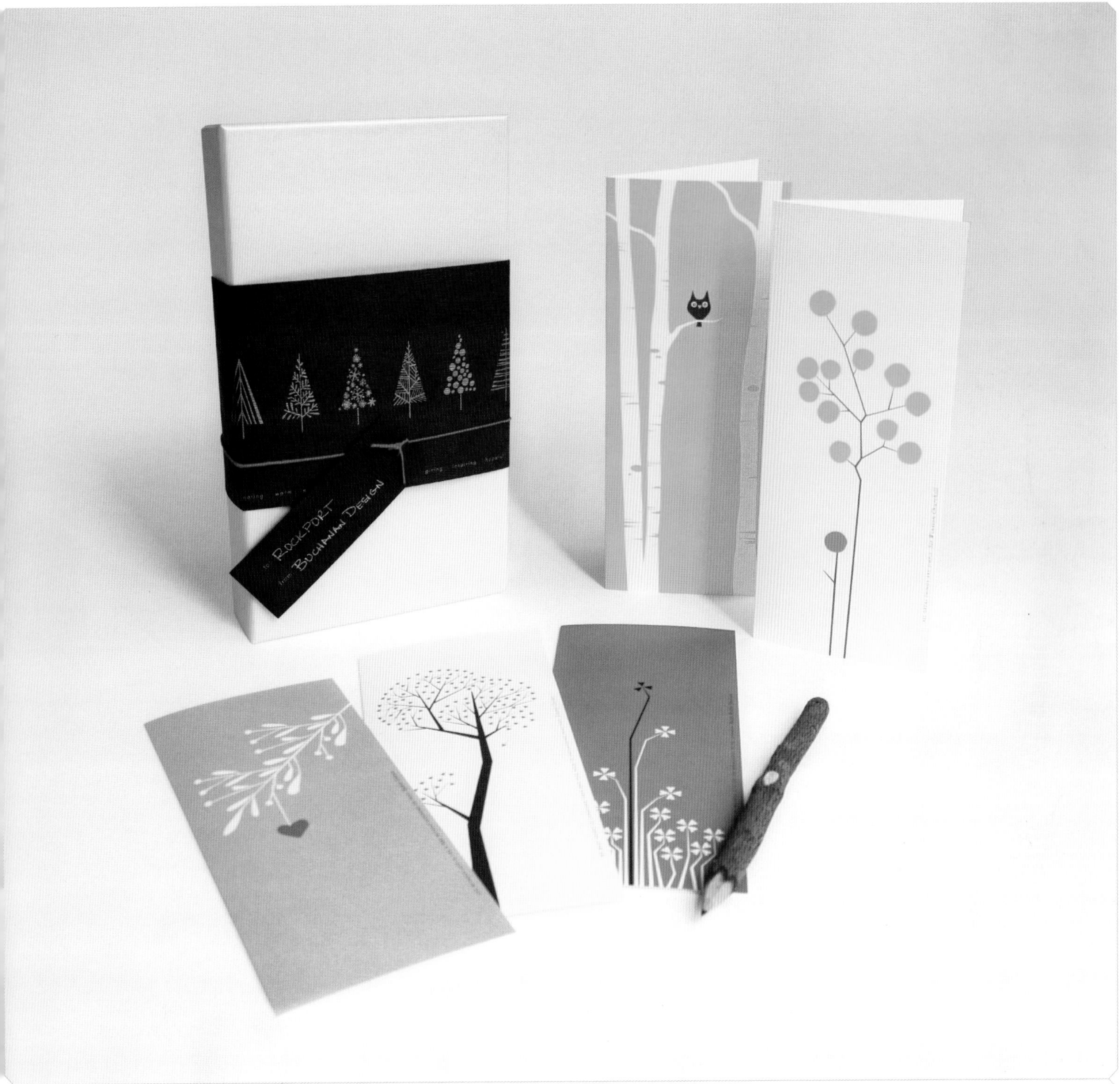
to ROCKPORT
from BUCHANAN DESIGN

BUCHANAN DESIGN
USA

0195

ARTISAN
BAKED BY HAND
BISCUITS
my favourite bear
BERRY GOOD BEAR
8 delicious bear-shaped natural all butter biscuits made with real BLACKCURRANT juice
my favourite bear
GO bananas BEAR
8 delicious bear-shaped natural all butter biscuits made with real BANANAS
ARTISAN
BAKED BY HAND
BISCUITS
my favourite bear
SNOW BEAR
8 delicious bear-shaped all butter WHITE CHOCOLATE biscuits with lots of YUMMY white chocolate
my favourite bear
SNOW BEAR

IRVING
UK

0196

TESCO
Plain
flour
Finer flour for successful baking
Perfect for pastries and biscuits
500g℮

P&W
UK
0197

P&W
UK
0198

TESCO
Strong
stoneground
100% wholemeal
bread flour
100%
Finer flour for successful baking
Perfect for cakes, biscuits and scones
500g℮

P&W
UK
0199

TESCO
Wholemeal
self raising
flour
Finer flour for successful baking
Perfect for cakes and scones
500g℮

P&W
UK
0200

ANTHEM WORLDWIDE
USA
0201

ANTHEM WORLDWIDE
USA
0202

ANTHEM WORLDWIDE
USA
0203

ANTHEM WORLDWIDE
USA
0204

ANTHEM WORLDWIDE
USA
0205

ANTHEM WORLDWIDE
USA
0206

ANTHEM WORLDWIDE
USA
0207

ANTHEM WORLDWIDE
USA
0208

ANTHEM WORLDWIDE
USA
0209

(oops)
Lost grape of Bordeaux discovered in Chile masquerading as a Merlot!
(oops)
There's a good reason the Merlot of Chile doesn't taste like other Merlot. Many are not entirely Merlot, but instead a blend of Merlot and a far more exotic grape: Carmenère, a.k.a., "The lost grape of Bordeaux." Carmenère had been transplanted from France over a century ago before the particularly nasty root louse threatened its very existence. Soon forgotten, the Carmenère grapes grew freely with Merlot grapes and were assumed to be Merlot until, suddenly in 1994, viticulturist Jean-Michel Boursiquit revealed the mistake. Oops!
Spicy splendor
With vivid red tones, aromas of red ripe fruit and chocolate, this spicy beauty is well-structured for a smooth finish. Fabulous with gamey meats and lamb.
Valle Central, Chile
100% CARMENÈRE
2005 Carmenère
(oops)
Lost grape of Bordeaux discovered in Chile masquerading as a Merlot!
(oops)
Voluptuous beauty
With its deep purple hues, this voluptuous beauty with coffee, blackberry, aromas of cherries and chocolate is divine with chicken, lamb, beef or spicy asian dishes.
Central Valley, Chile
84% CABERNET FRANC 16% CARMENÈRE
Cabernet Franc
(oops)
Lost grape of Bordeaux discovered in Chile masquerading as a Merlot!
(oops)
Cheeky little red.
With its ruby glow, this cheeky little red brimming with blackberry and oak flavors has an aroma of leather and vanilla. It's superb with red meat, or tomato pasta dishes.
Valle Central, Chile
70% CARMENÈRE 30% MERLOT
Carmenère Merlot

PEARLFISHER
UK

KASUBA DESIGN
USA

0211

TURNER DUCKWORTH
USA

0212

TURNER DUCKWORTH
USA

0213

TURNER DUCKWORTH
USA

214

TURNER DUCKWORTH
USA

0215

HORNALL ANDERSON DESIGN
USA
0216

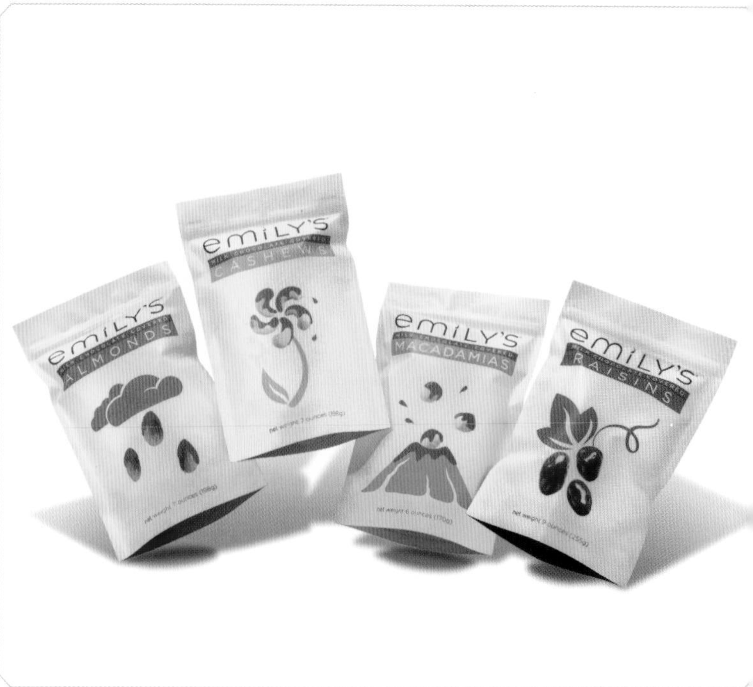

HORNALL ANDERSON DESIGN
USA
0217

P&W
UK
0218

P&W
UK
0219

TESCO
banana
flavoured
milk
• Controlled sugar
• No artificial flavours
• No artificial sweeteners
V
Vegetarian

P&W
UK

0220

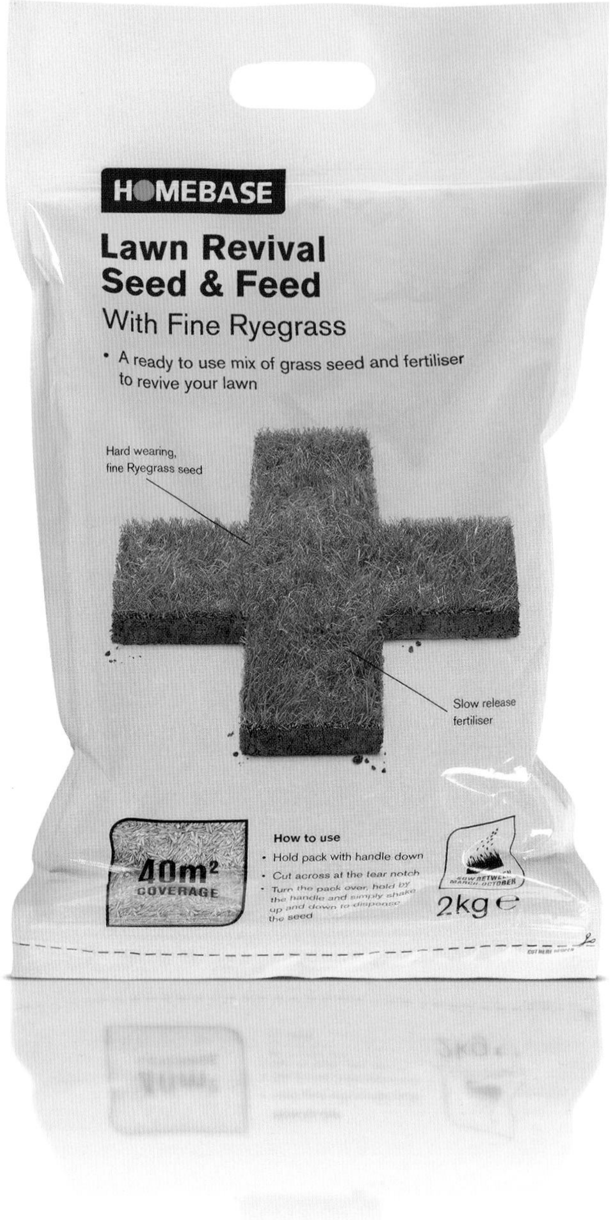
HOMEBASE
Lawn Revival
Seed & Feed
With Fine Ryegrass
• A ready to use mix of grass seed and fertiliser to revive your lawn
Hard wearing, fine Ryegrass seed
Slow release fertiliser
40m² COVERAGE
How to use
• Hold pack with handle down
• Cut across at the tear notch
• Turn the pack over, hold by the handle and simply shake up and down to dispense the seed
2kg ℮

TURNER DUCKWORTH
USA

0221

TURNER DUCKWORTH
USA

0222

TURNER DUCKWORTH
USA

0223

P&W
UK

0224

P&W
UK

0225

| TURNER DUCKWORTH | TURNER DUCKWORTH | TURNER DUCKWORTH |
| USA   0226 | USA   0227 | USA   0228 |

| JEN CADAM DESIGN | JEN CADAM DESIGN | JEN CADAM DESIGN |
| USA   0229 | USA   0230 | USA   0231 |

| JEN CADAM DESIGN | JEN CADAM DESIGN | JEN CADAM DESIGN |
| USA   0232 | USA   0233 | USA   0234 |

LEMLEY DESIGN COMPANY
USA

0235

Full tank
STAGE
2
Baby Fuel™
Pear & Quinoa Baby Food
USDA ORGANIC
Portable Purées
For Busy Babies!
ORGANIC
3x POUCHES
3-2.25 OZ (64g) POUCHES
NET WT 6.75 OZ (192g)
Butternut Squash Baby Food
Carrot Baby Food
STAGE
3

TURNSTYLE
USA

0236

YIYING LU DESIGN
AUSTRALIA

0237

YIYING LU DESIGN
AUSTRALIA

0238

YIYING LU DESIGN
AUSTRALIA

0239

TURNER DUCKWORTH
USA

0240

TURNER DUCKWORTH
USA
0241

TURNER DUCKWORTH
USA
0242

TURNER DUCKWORTH
USA
0243

TURNER DUCKWORTH
USA
0244

HERBS
·DE·
PROVENCE
FANTASTIQUE FOR ZHEE CHICKADEE
SELECTED SEASONINGS SUITABLE FOR THE SAVVY. PACKAGED AND PRESENTED BY BAKKEN CREATIVE CO.
ASIAN
RUB
RUB IT IN. GRILL IT UP.
FLEUR
·DE·
SEL
DAMN FANCY TABLE SALT

BAKKEN CREATIVE CO.
USA

0245

WINE THAT LOVES
ROASTED CHICKEN
WINE THAT LOVES
PASTA WITH TOMATO SAUCE

LIPPINCOTT
USA

0246

BLUE Q
USA

0247

BLUE Q
USA

0248

BLUE Q
USA

0249

BLUE Q
USA

0250

PINK BLUE BLACK & ORANGE CO., LTD.
THAILAND
0251

PINK BLUE BLACK & ORANGE CO., LTD.
THAILAND
0252

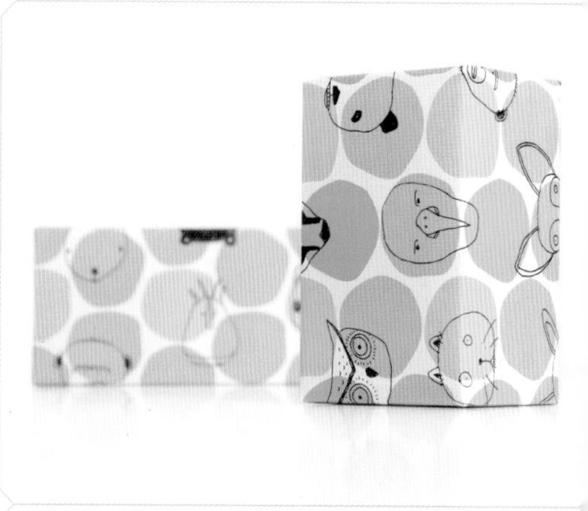

BBK STUDIO
USA
0253

MUCCA DESIGN CORP.
USA
0254

IRVING
UK
0255

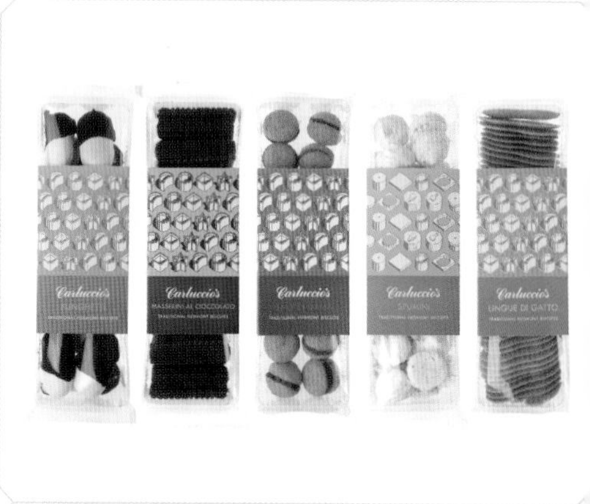

IRVING
UK
0256

RULE29
USA
0257

ROME & GOLD CREATIVE
USA
0258

ROME & GOLD CREATIVE
USA
0259

Duc de Praslin Belgium
Together Forever
HEARTS CREATE GOOD THINGS, ESPECIALLY LOVE
Be my LOVE!
Come & be my Valentine

PINK BLUE BLACK & ORANGE CO., LTD.
THAILAND

0260

Carluccio's
PANETTONE TRADIZIONALE
PANETTONE AL CAPPUCCINO
PANETTONE TRADIZIONALE
PANETTONE AL CAPPUCCINO

IRVING
UK

0261

THE REPUBLIC OF TEA
USA
0262

THE REPUBLIC OF TEA
USA
0263

THE REPUBLIC OF TEA
USA
0264

THE REPUBLIC OF TEA
USA
0265

THE REPUBLIC OF TEA
USA
0266

THE REPUBLIC OF TEA
USA
0267

THE REPUBLIC OF TEA
USA
0268

THE REPUBLIC OF TEA
USA
0269

THE REPUBLIC OF TEA
USA
0270

METHOD
USA
0271

METHOD
USA
0272

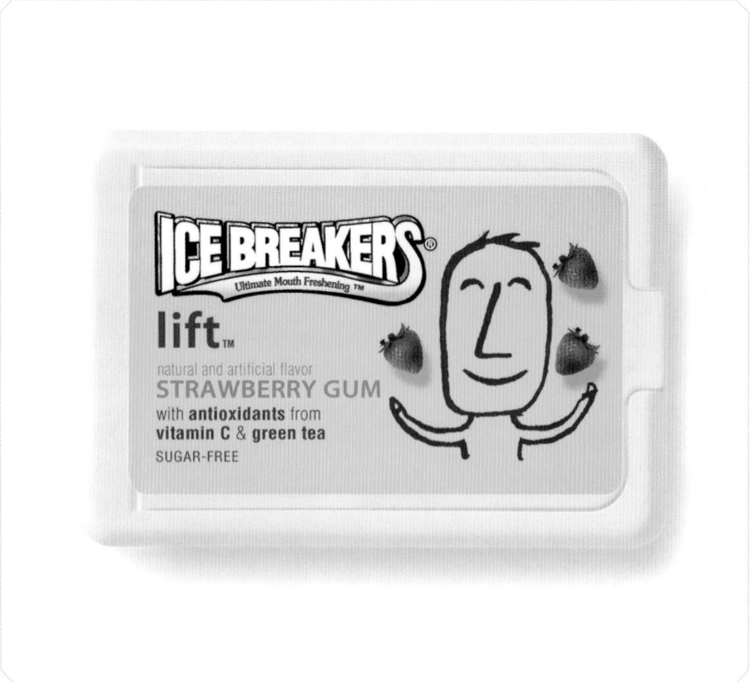

PEARLFISHER
USA
0273

PEARLFISHER
USA
0274

rice milk + mallow
method baby™
squeaky green™ hair + body wash
gentle, tear free +
99% natural +
pediatrician tested +
hypo-allergenic +
236mL (8 FL OZ)
rice milk + mallow
method baby™
squeaky green™ body lotion
gentle, non-irritating +
95% natural +
pediatrician tested +
hypo-allergenic +
236mL (8 FL OZ)
rice milk + mallow
method baby™
squeaky green™ diaper cream
gentle, non-irritating +
95% natural +
pediatrician tested +
hypo-allergenic +
153g (5.4 OZ)
rice milk + mallow
methodbaby™
squeaky green™ bubbly bath
gentle, tear free +
99% natural +
pediatrician tested +
hypo-allergenic +
295mL (10 FL OZ)

METHOD
USA

Get REAL
BARE NAKED
GARDEN
garden
ALL natural
Net Wt 2oz 59ml
NO fake crap
BODY SPLASH

BARE NAKED
Get REAL
garden

BLUE Q
USA

0276

BLUE Q
USA
0277

BLUE Q
USA
0278

TOKY BRANDING +DESIGN
USA
0279

TOKY BRANDING +DESIGN
USA
0280

CONCRETE DESIGN COMMUNICATIONS
CANADA
0281

CONCRETE DESIGN COMMUNICATIONS
CANADA
0282

ELEMENT
USA
0283

ELEMENT
USA
0284

ELEMENT
USA
0285

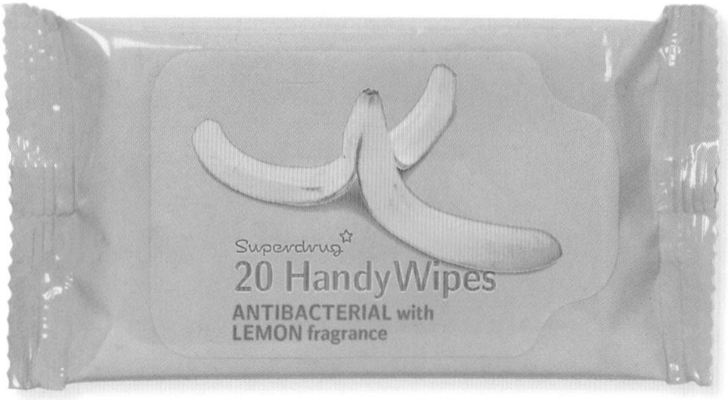

TURNER DUCKWORTH
USA
0286

TURNER DUCKWORTH
USA
0287

TURNER DUCKWORTH
USA
0288

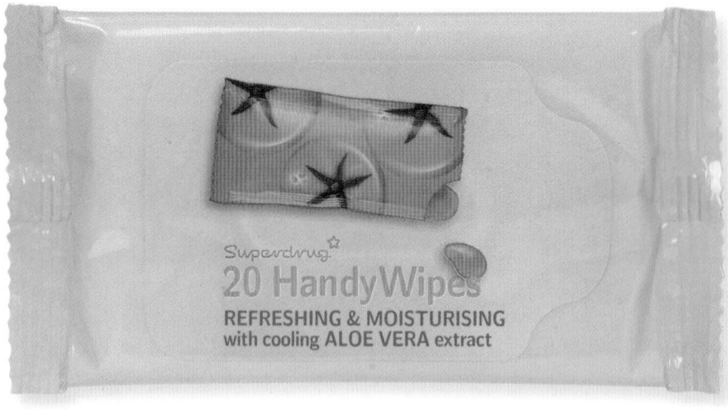

TURNER DUCKWORTH
USA
0289

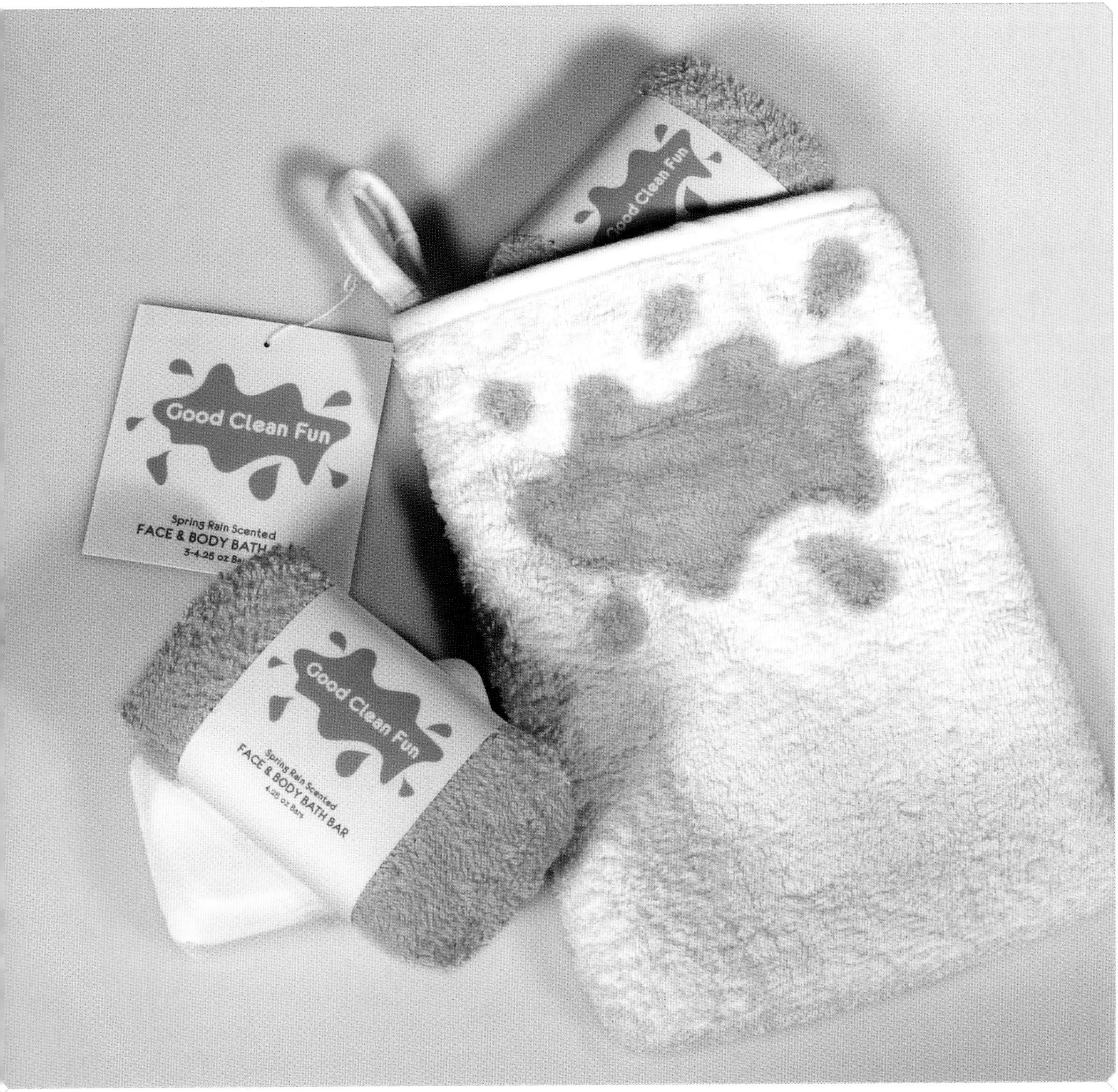
Good Clean Fun
Spring Rain Scented
FACE & BODY BATH BAR
Good Clean Fun
Spring Rain Scented
FACE & BODY BATH BAR
4.25 oz Bars
Good Clean Fun

CAZEE DESIGN
USA

0290

CD packaging KIT
CD packaging KIT
CHRONICLE BOOKS
turn your music and photo CDs into instant gifts!

CHRONICLE BOOKS
USA

0291

CHRONICLE BOOKS
USA

0292

CHRONICLE BOOKS
USA

0293

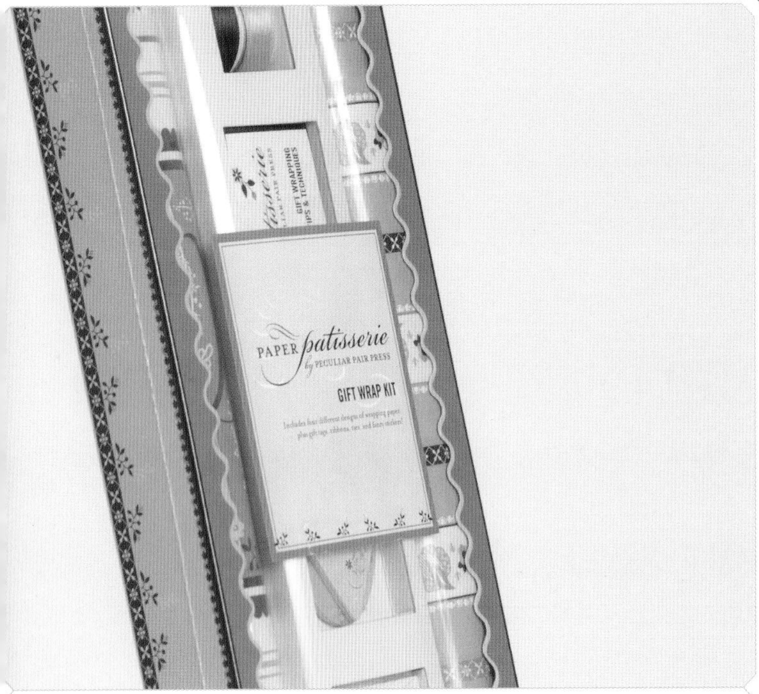

CHRONICLE BOOKS
USA

0294

HELLO! LUCKY
USA

0295

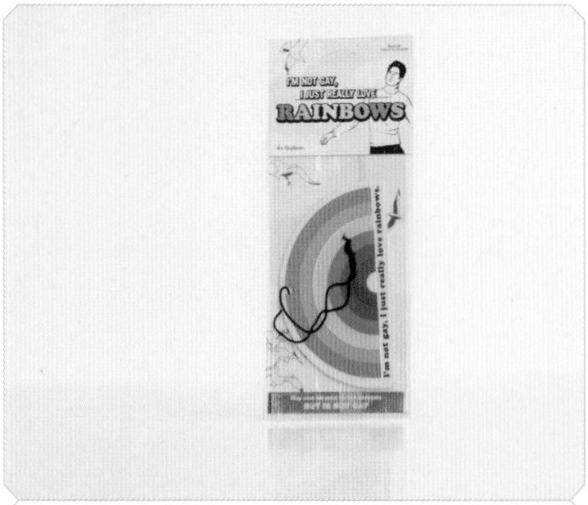

MODERN DOG DESIGN CO.
USA
0296

MODERN DOG DESIGN CO.
USA
0297

FRED / EASY ACES INC.
USA
0298

FRED / EASY ACES INC.
USA
0299

A L M PROJECT
USA
0300

212-BIG-BOLT
USA
0301

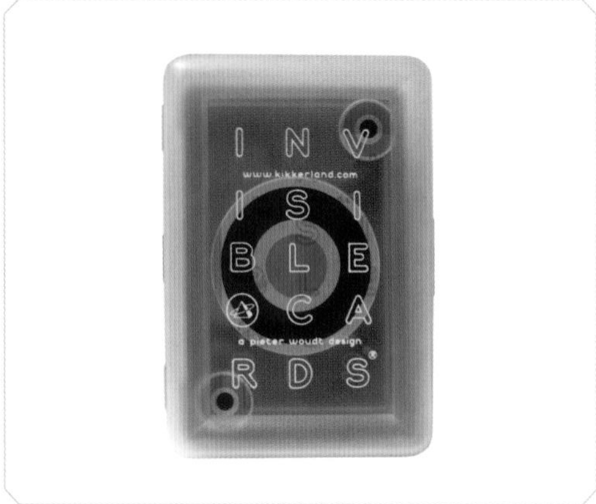

212-BIG-BOLT
USA
0302

BLUE Q
USA
0303

KENDALL ROSS
USA
0304

Sprinkles
CUPCAKE MIX
lemon
cupcake mix
makes 12
Handcrafted from the highest quality natural ingredients and topped with our signature modern dots, Sprinkles Cupcakes are a deliciously sophisticated take on an American classic. Since we opened the first cupcake bakery in Beverly Hills, our freshly baked treats have inspired long lines of devoted Hollywood stars and serious epicureans alike. Now, we bring our most popular flavors to the comfort of your own home with Sprinkles Cupcake Mixes.
NET WT. 16.2oz (460g)

A L M PROJECT
USA

0305

Chewy
CHEWING GUM
ASSISTANT ASSHOLE
It's Not PMS
It's You
8 ass-kicking pieces
YOU HAVE A GREAT ASS
Ham my ass!
CLASSIC MINT GUM
LET'S SHARE
EIGHT PIECES OF HOT CINNAMON GUM
Enough To Go Around

METHANE STUDIOS
USA

0306

CLIMATE CHANGE CHOCOLATE™
HELP SAVE THE EARTH. IT'S THE ONLY PLANET WITH CHOCOLATE ON IT!
Treat yourself while treating our planet a whole lot better.
CONTAINS: 3.5oz of feel good all natural premium DARK CHOCOLATE (55% cocoa)
plus TerraPass™ verified carbon offsets to balance one's average daily contribution to climate change.*
terrapass

BLOOMSBERRY & CO.
USA

0307

CLIMATE CHANGE CHOCOLATE™
HELP SAVE THE EARTH. IT'S THE ONLY PLANET WITH CHOCOLATE ON IT!
Treat yourself while treating our planet a whole lot better.
CONTAINS: 3.5oz of feel good all natural premium MILK CHOCOLATE (34% cocoa)
plus TerraPass™ verified carbon offsets to balance one's average daily contribution to climate change.*
terrapass

BLOOMSBERRY & CO.
USA

0308

CLIMATE CHANGE CHOCOLATE™
REDUCING YOUR CARBON FOOTPRINT JUST GOT A WHOLE LOT TASTIER!
Put your feet up and treat yourself while treating our planet better. After all, it's the only one with chocolate on it.
CONTAINS: 3.5oz of feel good all natural premium DARK CHOCOLATE (55% cocoa) plus TerraPass™ verified carbon offsets to balance one's average daily contribution to climate change.*
terrapass

BLOOMSBERRY & CO.
USA

0309

CLIMATE CHANGE CHOCOLATE™
REDUCING YOUR CARBON FOOTPRINT JUST GOT A WHOLE LOT TASTIER!
Put your feet up and treat yourself while treating our planet better. After all, it's the only one with chocolate on it.
CONTAINS: 3.5oz of feel good all natural premium MILK CHOCOLATE (34% cocoa) plus TerraPass™ verified carbon offsets to balance one's average daily contribution to climate change.*
terrapass

BLOOMSBERRY & CO.
USA

0310

BLOOMSBERRY & CO.
USA
0311

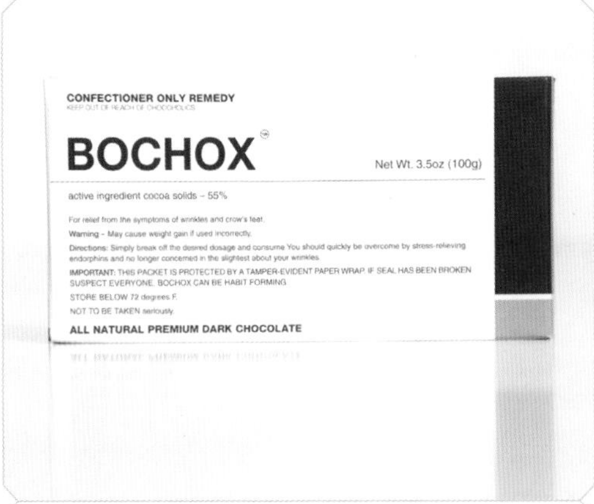

BLOOMSBERRY & CO.
USA
0312

BLOOMSBERRY & CO.
USA
0313

BLOOMSBERRY & CO.
USA
0314

BLOOMSBERRY & CO.
USA
0315

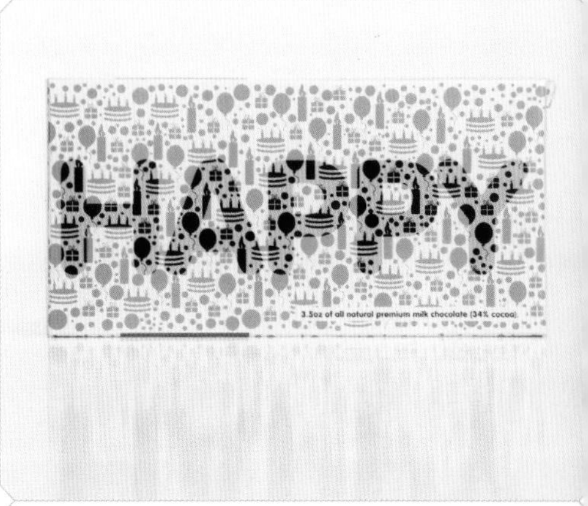

BLOOMSBERRY & CO.
USA
0316

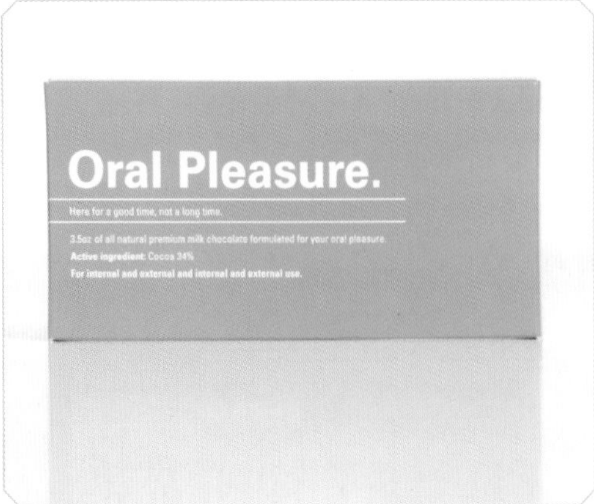

BLOOMSBERRY & CO.
USA
0317

BLOOMSBERRY & CO.
USA
0318

BLOOMSBERRY & CO.
USA
0319

BLOOMSBERRY & CO.
USA

0320

Think chocolate™
10 bars (3.5oz Per Bar)
Net Weight 35oz (1000g)
All Natural

PLEASE
TO
ENJOY
DELUXE
EXOTIC
华贵
珍奇
MISO
PRETTY
Glitter
BALL
厂家
清馨来自
PEARLY
RECIPE

MISO PRETTY
DELUXE
华贵
MISO
PRETTY
Glitter BALL

BLUE Q
USA

0321

BLUE Q
USA
0322

BLUE Q
USA
0323

BLUE Q
USA
0324

BLUE Q
USA
0325

FRED / EASY ACES INC.
USA
0326

FRED / EASY ACES INC.
USA
0327

FRED / EASY ACES INC.
USA
0328

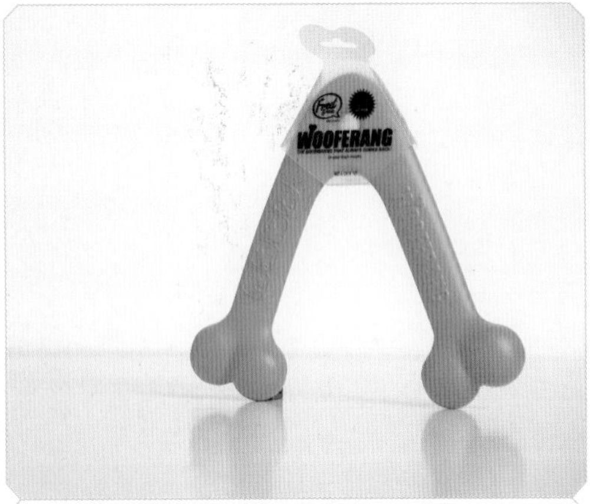

FRED / EASY ACES INC.
USA
0329

FRED / EASY ACES INC.
USA
0330

FRED / EASY ACES INC.
USA
0331

FRED / EASY ACES INC.
USA
0332

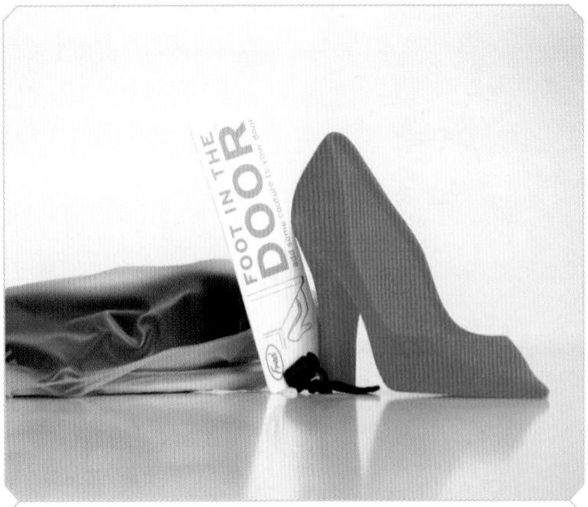

FRED / EASY ACES INC.
USA
0333

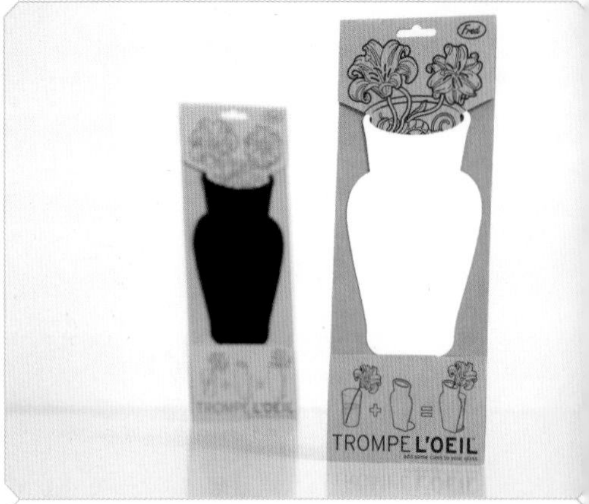

FRED / EASY ACES INC.
USA
0334

SMARTCOOKIE
ask the cookie
shake it twice
answer revealed
follow advice
Inscrutable Fred

FRED / EASY ACES INC.
USA

0335

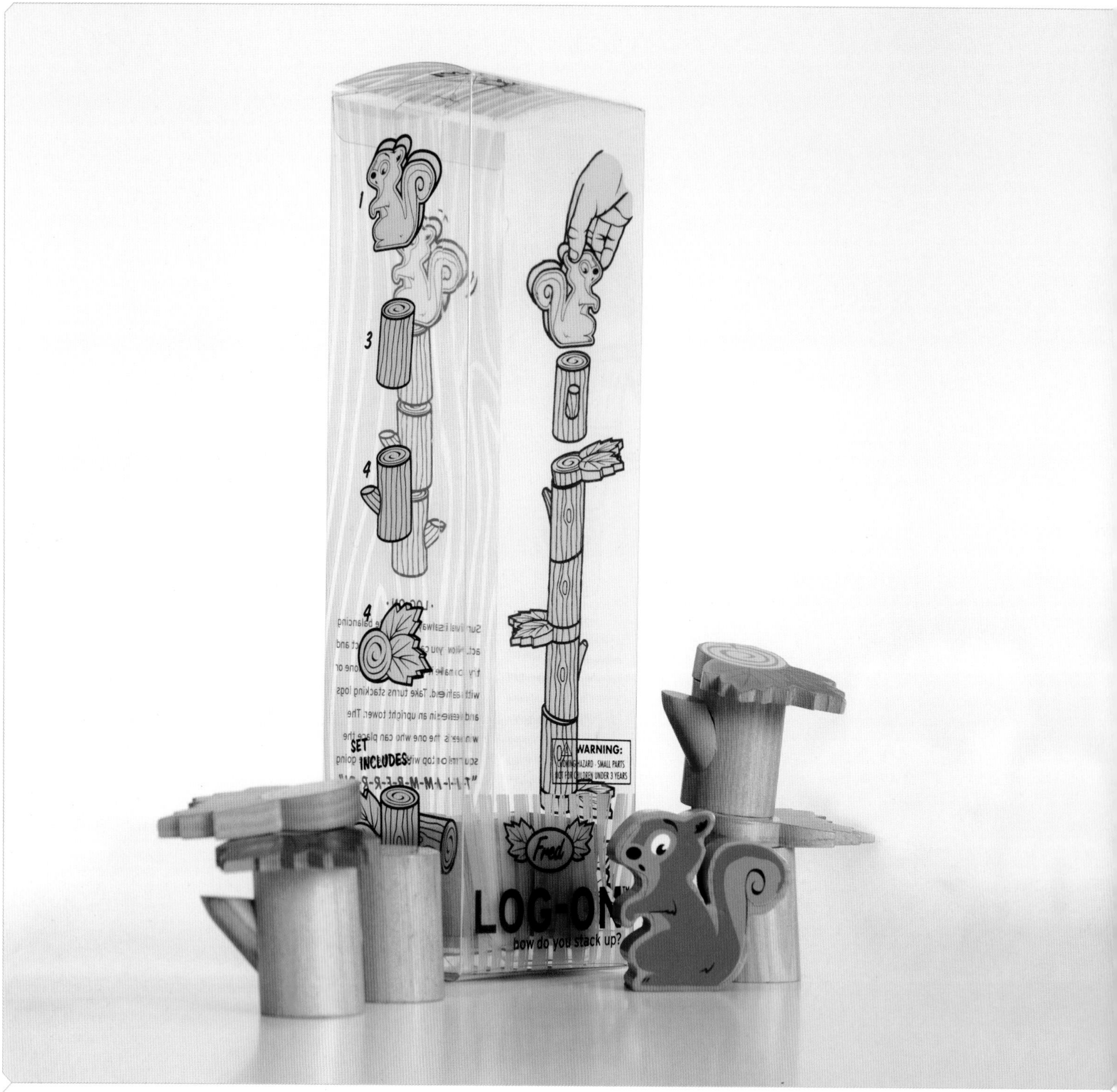
Fred
LOG-ON™
how do you stack up?
WARNING:
SET INCLUDES:

FRED / EASY ACES INC.
USA

0336

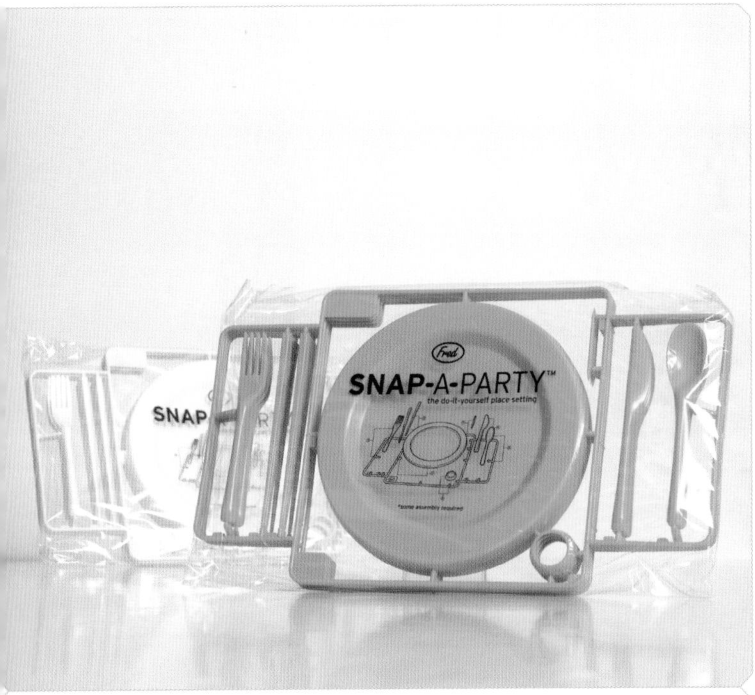

FRED / EASY ACES INC.
USA
0337

FRED / EASY ACES INC.
USA
0338

FRED / EASY ACES INC.
USA
0339

FRED / EASY ACES INC.
USA
0340

FRED / EASY ACES INC.
USA
0341

FRED / EASY ACES INC.
USA
0342

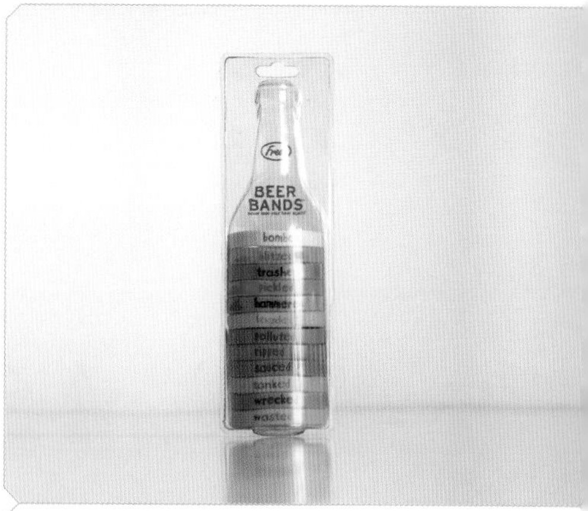

FRED / EASY ACES INC.
USA
0343

FRED / EASY ACES INC.
USA
0344

FRED / EASY ACES INC.
USA
0345

OXIDE DESIGN CO.
USA
0346

OXIDE DESIGN CO.
USA
0347

WALLACE CHURCH, INC.
USA
0348

WALLACE CHURCH, INC.
USA
0349

CIAO BELLA
STRAWBERRY CHARDONNAY • FRAGOLE IN CHARDONNAY
SORBETTO
CIAO BELLA GELATO CO., INC.
GELATO
MINT CHOCOLATE CHIP
FRESCA CON CIOCCOLATO
CHOCOLATE CHIP
1 PINT
(473 mL)

WALLACE CHURCH, INC.
USA

0350

Dots & Jots
By Denyse Schmidt
NOTECARD BOOK

CHRONICLE BOOKS
USA

0351

CHRONICLE BOOKS
USA
0352

DESIGN BRIDGE
UK
0353

DESIGN BRIDGE
UK
0354

DESIGN BRIDGE
UK
0355

word!

OUTSPOKEN
BOLD
SAVVY
CHARISMATIC
ENGAGING
SHREWD
VIVID
PROVOCATIVE
INTRIGUING
DARING
HIP
SPIRITED

GUZMAN Y GOMEZ
Taquería
AND SKINLESS FREE
RANGE CHICKENS. WE
USE ONLY OCEAN FRESH
SEAFOOD, FRESH GARDEN
VEGETABLES AND ARE
EQUALLY STRICT WITH
ALL OUR SEASONINGS.
THIS IS FOOD
TRADITIONALLY HIGH
IN NUTRIENTS, PACKED
WITH VITAMINS AND
LOW IN FAT. WITH ONLY
THE EXCEPTION OF
OUR TOTOPOS, NOTHING
IS COOKED IN OIL.
Auténtico, Mexicano, Delicioso

THE CREATIVE METHOD
AUSTRALIA

0356

EMMI
UK
0357

THE CREATIVE METHOD
AUSTRALIA
0358

THE CREATIVE METHOD
AUSTRALIA
0359

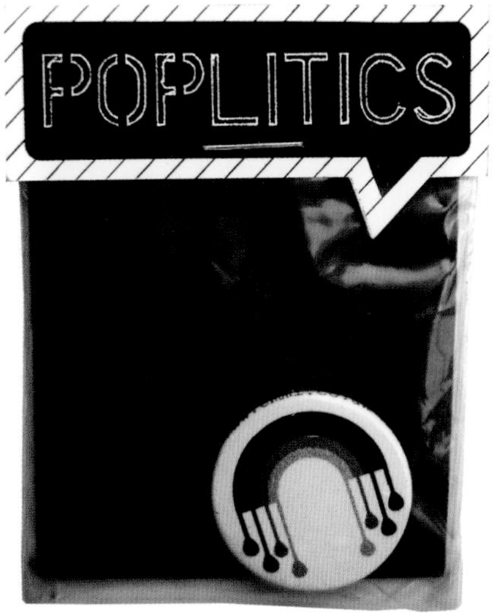

EMMI
UK
0360

ZION GRAPHICS
SWEDEN

0361

MILTON GLASER, INC.
USA

0362

MILTON GLASER, INC.
USA

0363

MILTON GLASER, INC.
USA

0364

LOCAL
1
BROOKLYN
BROOKLYN
LOCAL
1
100% BOTTLE
REFERMENTED ALE

MILTON GLASER, INC.
USA

0365

Douglas Gordon's
The
VANITY
of
Allegory

SAGMEISTER INC.
USA

0366

SYNTHETIC INFATUATION
USA
0367

LA SUPERAGENCIA C
VENEZUELA
0368

CHRONICLE BOOKS
USA
0369

CONCRETE DESIGN COMMUNICATIONS
CANADA
0370

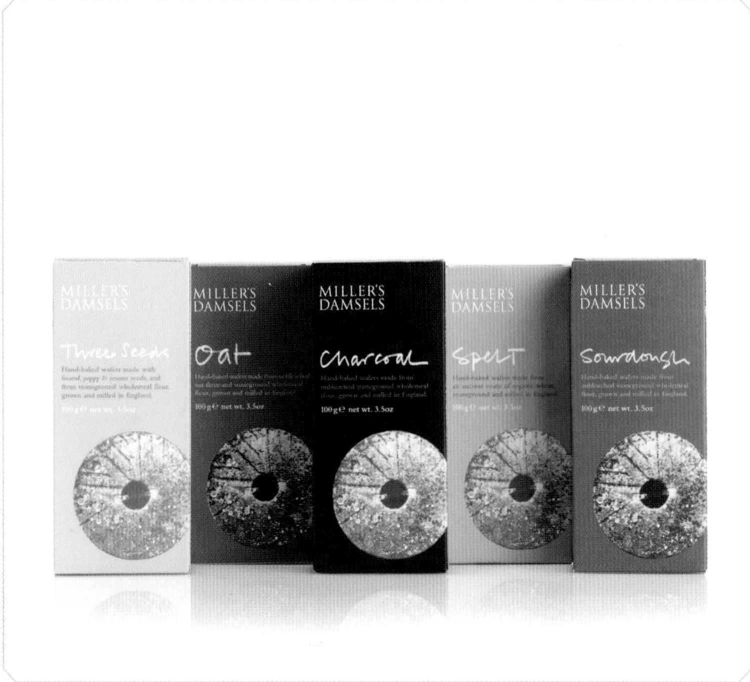

IRVING
UK
0371

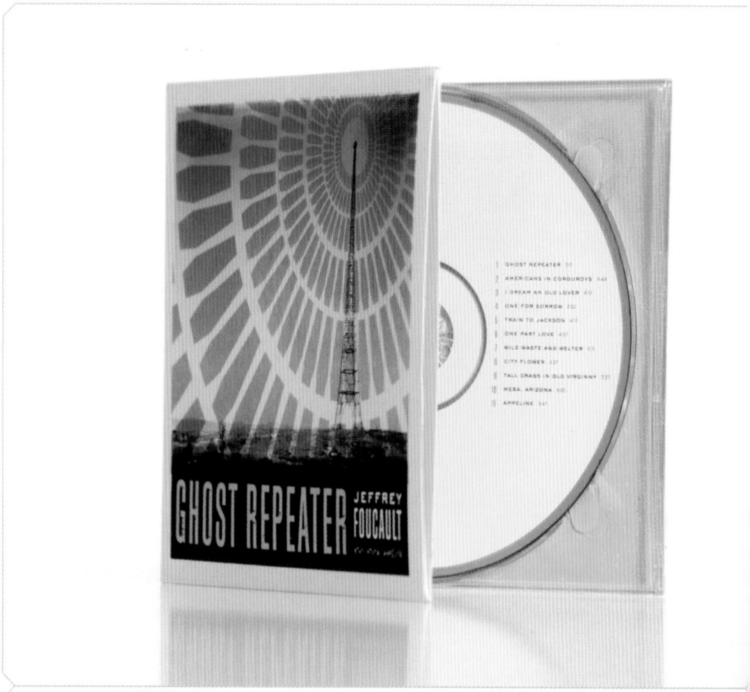

THE DECODER RING DESIGN CONCERN
USA
0372

ANTONIN FERLA
UK
0373

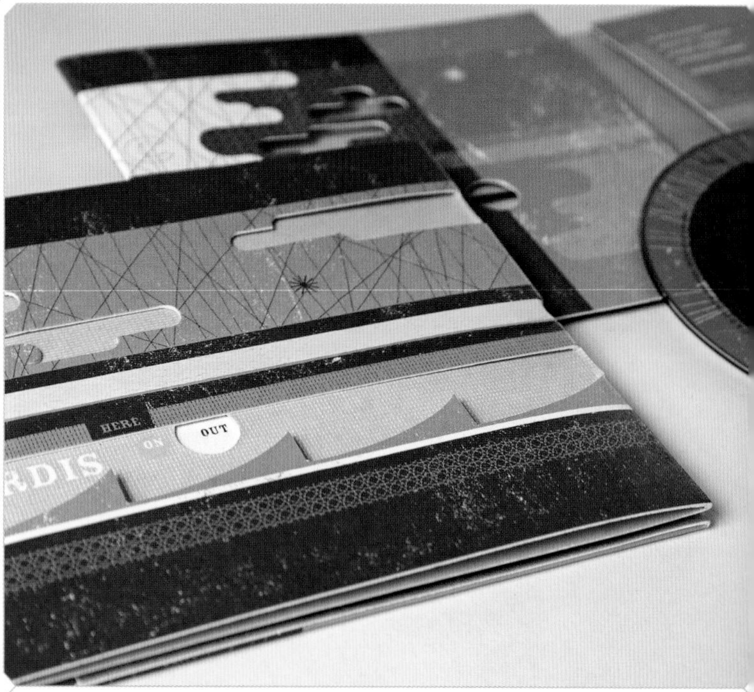

CHEN DESIGN ASSOCIATES
USA
0374

MILLER'S DAMSELS
Three Seeds
Hand-baked wafers made with linseed, poppy & sesame seeds, and from 100% stoneground wholemeal flour, grown and milled in England.
MILLER'S DAMSELS
Oat
Hand-baked wafers made from unbleached oat flour and stoneground wholemeal flour, grown and milled in England.
MILLER'S DAMSELS
Sourdough
Hand-baked wafers made from 100% unbleached stoneground wholemeal flour, grown and milled in England.
MILLER'S DAMSELS
Rye
Wheat-free hand-baked wafers made from 100% unbleached stoneground rye flour, grown and milled in England.
MILLER'S DAMSELS
Wheat
Hand-baked wafers made from 100% unbleached stoneground wholemeal flour, grown and milled in England.
MILLER'S DAMSELS
Charcoal
Hand-baked wafers made from 100% unbleached stoneground wholemeal flour, grown and milled in England.

IRVING
UK

0375

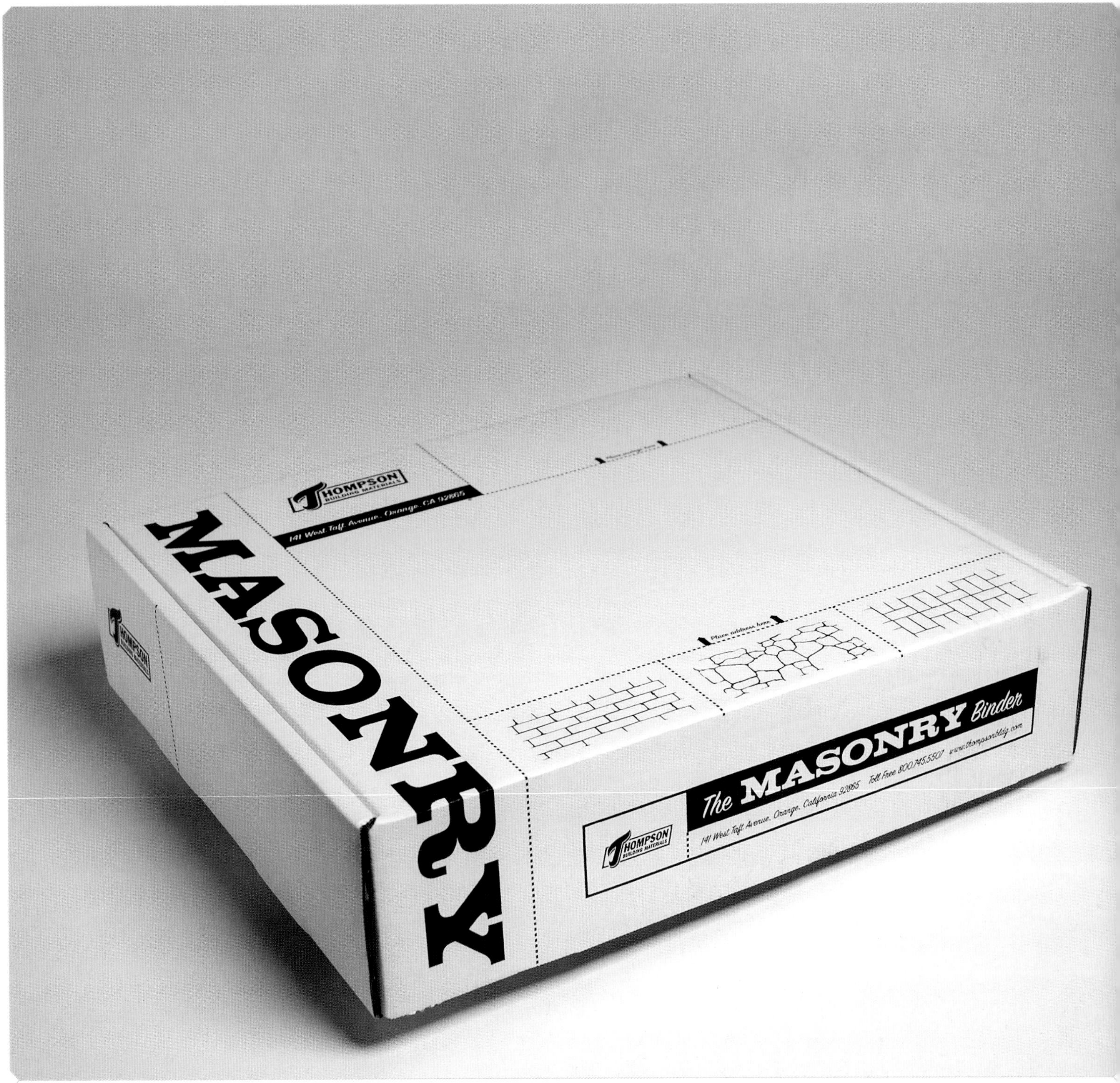
MASONRY
THOMPSON
BUILDING MATERIALS
141 West Taft Avenue, Orange, CA 92865
The MASONRY Binder
141 West Taft Avenue, Orange, California 92865

CONOVER
USA

0376

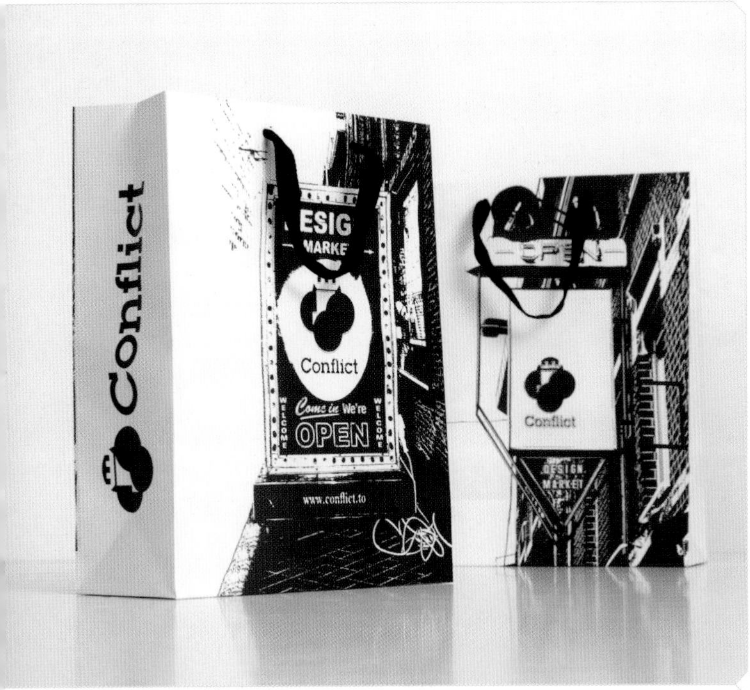

BOY BASTIAENS
THE NETHERLANDS
0377

LESS RAIN
JAPAN
0378

BLACKBOOKS
USA
0379

SATELLITES MISTAKEN FOR STARS
AUSTRIA
0380

J. KENNETH ROTHERMICH
USA
0381

KEN-TSAI LEE DESIGN STUDIO
TAIWAN
0382

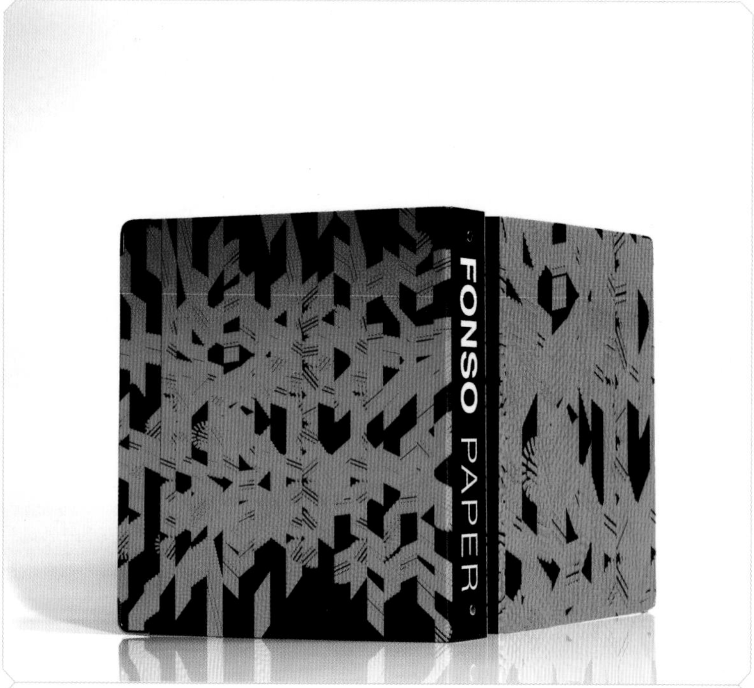

KEN-TSAI LEE DESIGN STUDIO
TAIWAN
0383

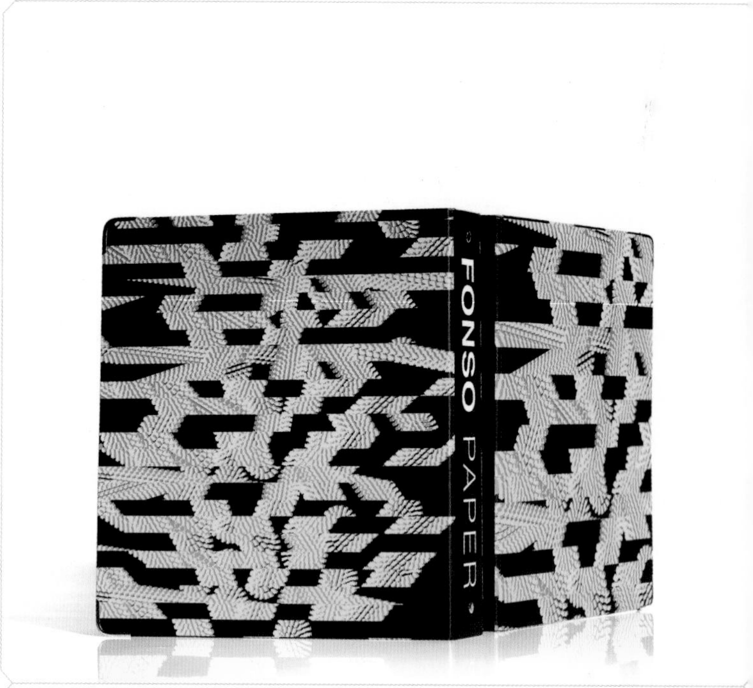

KEN-TSAI LEE DESIGN STUDIO
TAIWAN
0384

AXE LAB
AXE LAB
MASSAGE OIL
4 FL OZ 118mL
AXE LAB
EAU DE TOILETTE

JOED DESIGN INC.
USA

0385

CHR JINGLES | SPRING07
FEATURING KIIS SUITE 2007
REELWORLD
CHR JINGLES | SPRING
REELWORLD
REELWORLD

TURNSTYLE
USA

0386

TURNSTYLE
USA

0387

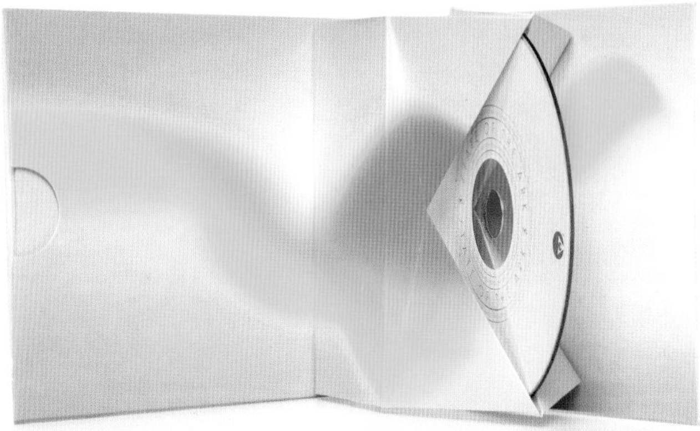

ZION GRAPHICS
SWEDEN

0388

SAYLES GRAPHIC DESIGN
USA

0389

TURNSTYLE
USA

0390

Goji Force Green Tea
fullyloadedtea

SUBPLOT DESIGN INC.
CANADA
0391

When did we start to say "and the award goes to" instead of "and the winner is"? Just seems so weak. Shameless Berries Fruit Tea is more likely to jump up on stage, grab the mic, and give a shout-out to its peeps. Mad love. We out!
fullyloadedtea
www.fullyloadedtea.com

SUBPLOT DESIGN INC.
CANADA
0392

Why all the touchy-feely corporate team-building exercises with people you see all day, every day? Tomato Rebel Black Tea says "enough!" Work hard and play even harder. But on your own terms. Power to the people!
WHOLE LEAF BLACK TEA
fullyloadedtea

SUBPLOT DESIGN INC.
CANADA
0393

You know what the problem with "whatever" is? It's got no guts. It just doesn't take a stand. Well, our Bulletproof Currant Fruit Tea doesn't have that problem. It puts it all out there and says, "love me or hate me, but I am me!"
CAFFEINE FREE WHOLE FRUIT TEA
fullyloadedtea

SUBPLOT DESIGN INC.
CANADA
0394

Hundreds of channels but nothing's on. Another cop show from the people who brought us that other cop show. Citrus Spiked Black Tea says "get out there!" You are your own real-life drama, not a flat-screen substitute. Tune out and turn on!
WHOLE LEAF BLACK TEA
fullyloadedtea

SUBPLOT DESIGN INC.
CANADA
0395

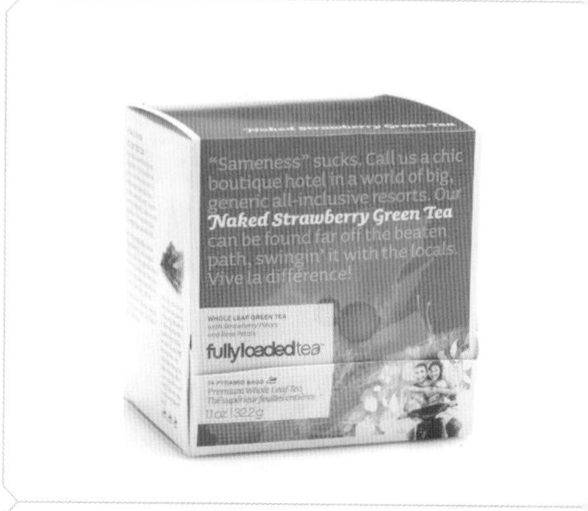
"Sameness" sucks. Call us a chic boutique hotel in a world of big, generic all-inclusive resorts. Our Naked Strawberry Green Tea can be found far off the beaten path, swingin' it with the locals. Vive la différence!
WHOLE LEAF GREEN TEA
fullyloadedtea

SUBPLOT DESIGN INC.
CANADA
0396

What's that? Recycled beats and faux-punk-pop? Hasn't anyone got anything original anymore? Sweet Addiction Fruit Tea is the kind of tune that works in the stadium or on the acoustic. Authentic and classic. Belt it out!
CAFFEINE FREE WHOLE FRUIT TEA
fullyloadedtea

SUBPLOT DESIGN INC.
CANADA
0397

When did we start to say "and the award goes to" instead of "and the winner is"? Just seems so weak. Shameless Berries Fruit Tea is more likely to jump up on stage, grab the mic, and give a shout-out to its peeps. Mad love. We out!
CAFFEINE FREE WHOLE FRUIT TEA
fullyloadedtea

SUBPLOT DESIGN INC.
CANADA
0398

Cucumber Mojo White Tea
WHOLE LEAF WHITE TEA
fullyloadedtea

SUBPLOT DESIGN INC.
CANADA
0399

fullyloadedtea
You know what the problem with "whatever" is? It's got no guts. It just doesn't take a stand. Well, our Bulletproof Currant Fruit Tea doesn't have that problem. It puts it all out there and says, "love me or hate me, but I am me!"
CAFFEINE FREE WHOLE FRUIT TEA with Blackcurrants, Cranberries, Lingonberries and Rosehips
fullyloadedtea
14 PYRAMID BAGS
Premium Whole Fruit Tea
Thé supérieur fruits entiers
1.0 oz | 28 g
Hundreds of channels but nothing's on. Another cop show from the people who brought us that other cop show. Citrus Spiked Black Tea says "get out there!" You are your own real-life drama, not a flat-screen substitute. Tune out and turn on!
WHOLE LEAF BLACK TEA with Orange Peel and Orange Flowers
fullyloadedtea
14 PYRAMID BAGS
Premium Whole Leaf Tea
Thé supérieur feuilles entières
1.0 oz | 28 g
"Sameness" sucks. Call us a chic boutique hotel in a world of big, generic all-inclusive resorts. Our Naked Strawberry Green Tea can be found far off the beaten path, swingin' it with the locals. Vive la différence!
WHOLE LEAF GREEN TEA with Strawberry Pieces and Rose Petals
fullyloadedtea
14 PYRAMID BAGS
Premium Whole Leaf Tea
Thé supérieur feuilles entières
1.1 oz | 32.2 g
What's that? Recycled beats and faux-punk-pop? Hasn't anyone got anything original anymore? Sweet Addiction Fruit Tea is the kind of tune that works in the stadium or on the acoustic. Authentic and classic. Belt it out!
CAFFEINE FREE WHOLE FRUIT TEA with Rosehips, Blackcurrant Leaves and Summer Flowers
fullyloadedtea
14 PYRAMID BAGS
Premium Whole Fruit Tea
Thé supérieur fruits entiers
1.0 oz | 28 g
When did we start to say "and the award goes to" instead of "and the winner is"? Just seems so weak. Shameless Berries Fruit Tea is more likely to jump up on stage, grab the mic, and give a shout-out to its peeps. Mad love. We out!
CAFFEINE FREE WHOLE FRUIT TEA with Goji Berries, Blueberries and Seabuckthorn Berries
fullyloadedtea
14 PYRAMID BAGS
Premium Whole Fruit Tea
Thé supérieur fruits entiers
1.0 oz | 28 g
We've all been there. Trying to ignore the muzak on the elevator. I mean, what's with that? Even our Goji Force Green Tea has to cover its ears. Really. Let's hear something with a bit more oomph. Rock it up or turn it off!
WHOLE LEAF GREEN TEA with Goji Berries and Lingonberries
fullyloadedtea
14 PYRAMID BAGS
Premium Whole Leaf Tea
Thé supérieur feuilles entières
1.1 oz | 32.2 g

SUBPLOT DESIGN INC.
CANADA

0400

NO. 0001
OR
ORIGINAL
JASON MARKK
PREMIUM SNEAKER SOLUTION
JASONMARKK.COM
jasonmarkk
JASON MARKK
PREMIUM SNEAKER SOLUTION
A DAILY RITUAL
NO. 0001
OR
ORIGINAL
236 ML 8 FL. OZ
CAUTION: KEEP OUT OF REACH OF CHILDREN. EYE IRRITANT. SEE BACK LABEL FOR ADDITIONAL PRECAUTIONARY AND PRACTICAL TREATMENT STATEMENTS.

JASON MARKK INC.
USA

0401

WALLACE CHURCH, INC.
USA

0402

ELEVATOR
CROATIA

0403

ELEVATOR
CROATIA

0404

ELEVATOR
CROATIA

0405

BFG
USA
0406

BFG
USA
0407

BFG
USA
0408

BAKKEN CREATIVE CO.
USA
0409

BAKKEN CREATIVE CO.
USA
0410

BAKKEN CREATIVE CO.
USA
0411

HOMART
USA
0412

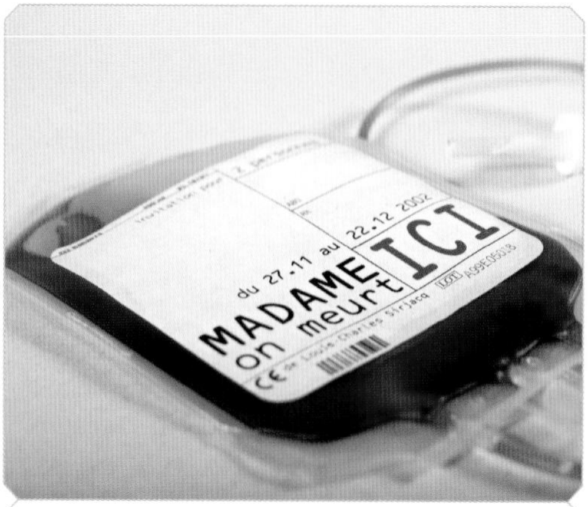

AUDE FERLA
UK
0413

COMPASS DESIGN
USA
0414

SAGMEISTER INC.
USA

0415

FIREBELLY DESIGN
When you believe strongly in the idea, people listen. When you feel passionately about the message, people trust you. And when you go with your gut, people see that and respect you for it. That's our definition of successful design and the type of dialogue we've always enjoyed with our clients and their audiences. Our clients entrust us with their brands, products and campaigns in turn we provide them with integrity, understanding and design that not only wins awards but hearts as well. We think that's a pretty fair trade.
FIREBELLY
FIREBELLY DESIGN
GOOD DESIGN FOR GOOD REASON
DAWN HANCOCK creative director

FIREBELLY DESIGN
USA

0416

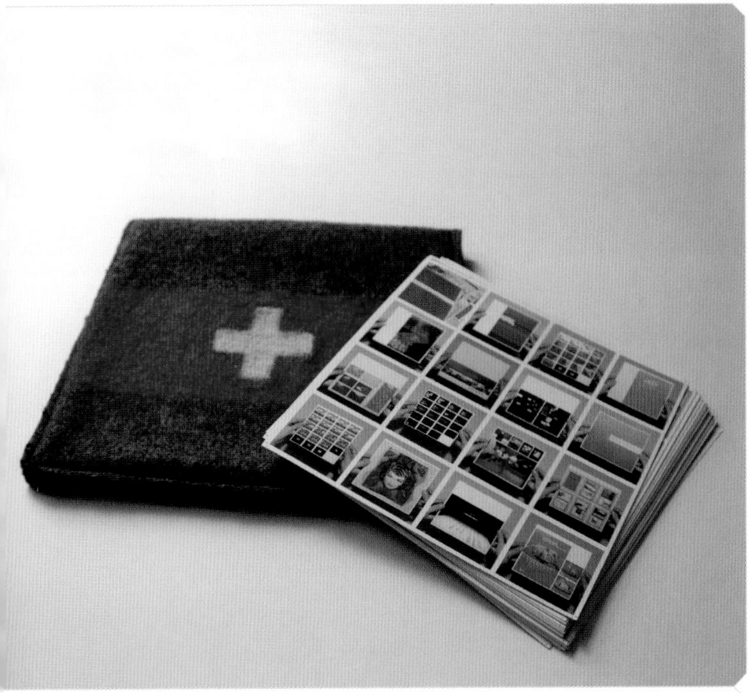

ANTONIN FERLA
UK

0417

MUCCA DESIGN CORP.
USA

0418

EMMI
UK

0419

EMMI
UK

0420

OBJECT 9
USA
0421

GIORGIO DAVANZO DESIGN
USA
0422

GIORGIO DAVANZO DESIGN
USA
0423

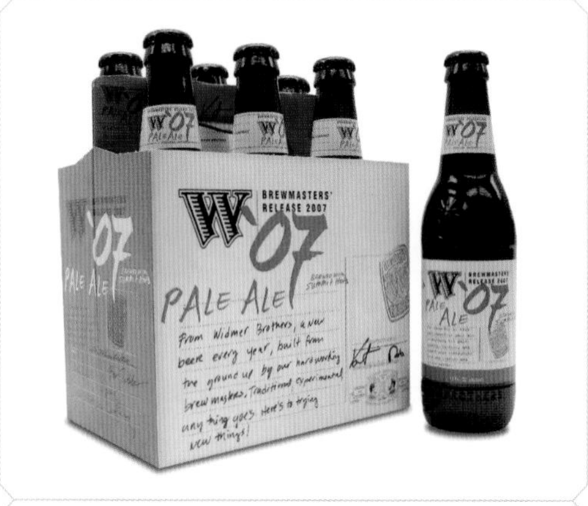

HORNALL ANDERSON DESIGN
USA
0424

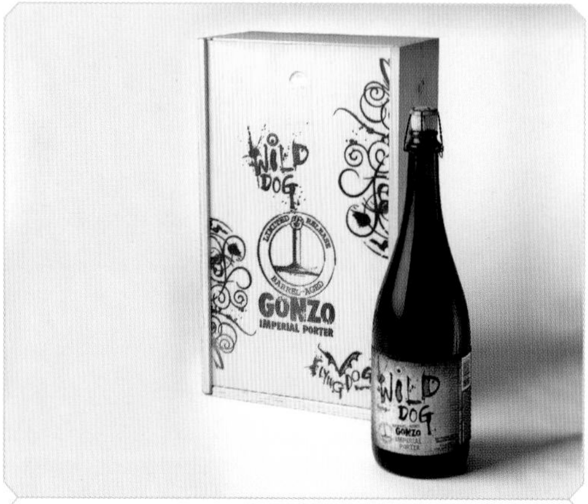

OBJECT 9
USA
0425

RANDY MOSHER DESIGN
USA
0426

CORNERSTONE STRATEGIC BRANDING
USA
0427

BEARBEAR CREATIVE
USA
0428

COMPASS DESIGN
USA
0429

BLANCO
BAMBARRIA
TEQUILA
100% BLUE AGAVE
GENUINE MEXICAN
TEQUILA. ESTATE
BOTTLED IN JALISCO
IMPORTED
38%VOL

P&W
UK

0430

THE DECODER RING DESIGN CONCERN
USA

0431

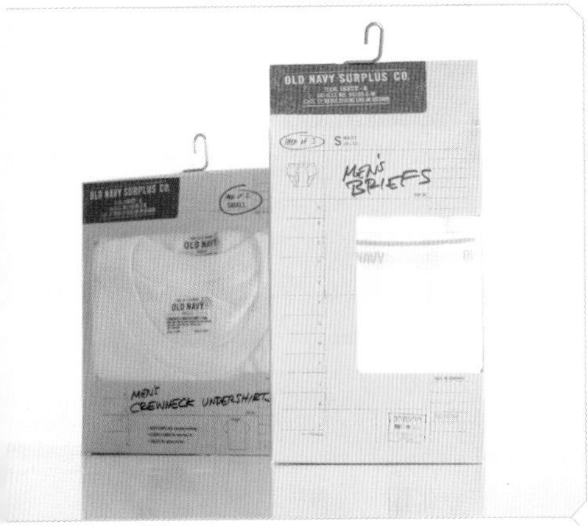

OLD NAVY
USA
0432

BONBON LONDON
UK
0433

BONBON LONDON
UK
0434

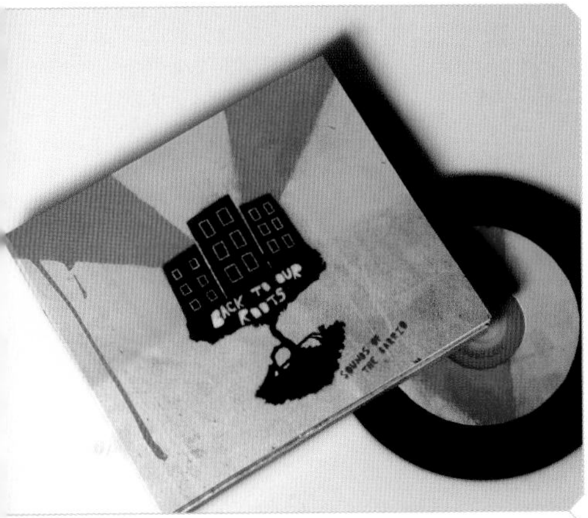

GNOME
USA
0435

SEGURA INC.
USA
0436

SEGURA INC.
USA
0437

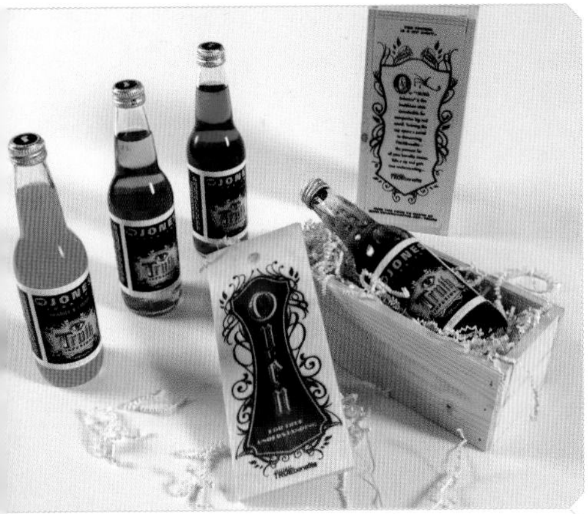

WOLKEN COMMUNICA
USA
0438

CONOVER
USA
0439

AUDE FERLA
UK
0440

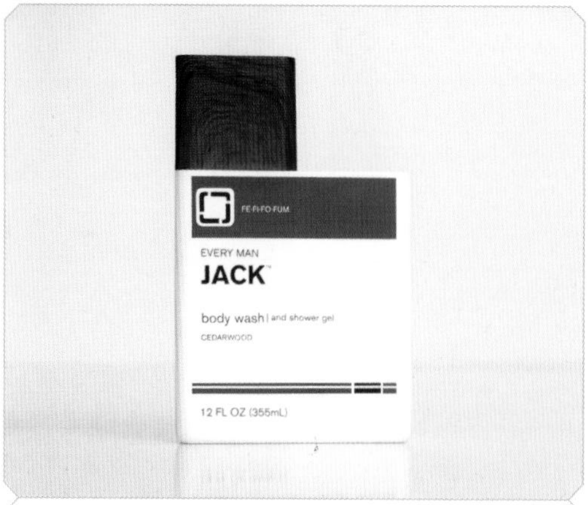

Y STUDIOS LLC
USA
0441

Y STUDIOS LLC
USA
0442

Y STUDIOS LLC
USA
0443

Y STUDIOS LLC
USA
0444

Y STUDIOS LLC
USA
0445

Y STUDIOS LLC
USA
0446

Y STUDIOS LLC
USA
0447

Y STUDIOS LLC
USA
0448

Y STUDIOS LLC
USA
0449

NEVER A DULL BOY.
EVERY MAN JACK™
face scrub | pre-shave
FRAGRANCE FREE
FL OZ (150mL)

Y STUDIOS LLC
USA

0450

WAKE UP CALL
EVERY MAN JACK™
body wash | and shower scrub
CITRUS SCRUB
12 FL OZ (355mL)

Y STUDIOS LLC
USA

0451

GO BIG
USA
0452

MICHAEL COURTNEY DESIGN, INC.
USA
0453

MICHAEL COURTNEY DESIGN, INC.
USA
0454

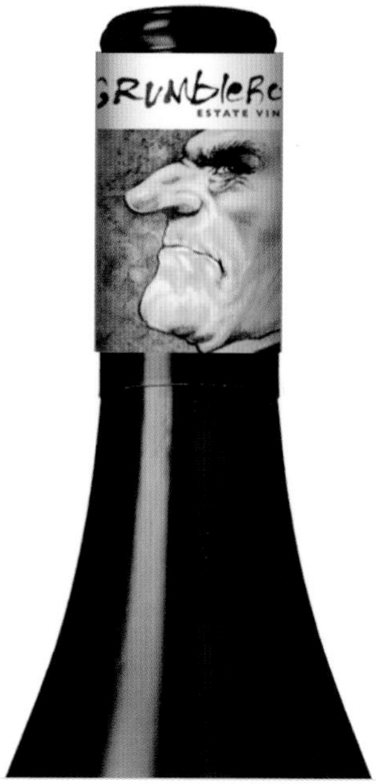

THE CREATIVE METHOD
AUSTRALIA
0455

ohlala!

LUXURIOUS
CLASSIC
DIVINE
RADIANT
DAZZLING
CLASSY
ELEGANT
EXQUISITE
FINE
LOVELY
SPLENDID
LUSCIOUS

TRADE LOLLIA MARK
RELAX
FOAMING BATH
Lavender & Honey | Evening Relaxation
LOLLIA
RELAX

BURWELL IND. INC.
USA

0456

BURWELL IND. INC.
USA

0457

BURWELL IND. INC.
USA

0458

BURWELL IND. INC.
USA

0459

BURWELL IND. INC.
USA

0460

BURWELL IND. INC.
USA

0461

BURWELL IND. INC.
USA

0462

BURWELL IND. INC.
USA

0463

BURWELL IND. INC.
USA

0464

TRADE LOLLIA MARK
triplets
No. 035
LOLLIA
1000/flowers
INSPIRE
SHEA BUTTER HANDCREME
finest perfumes X 3
1.5 oz X 3

BURWELL IND. INC.
USA

0465

BURWELL IND. INC.
USA

0466

NIEDERMEIER DESIGN
USA

0467

NIEDERMEIER DESIGN
USA

0468

NIEDERMEIER DESIGN
USA

0469

NIEDERMEIER DESIGN
USA

0470

ABSOLUTE ZERO DEGREES
UK

0471

ABSOLUTE ZERO DEGREES
UK

0472

ABSOLUTE ZERO DEGREES
UK

0473

ABSOLUTE ZERO DEGREES
UK

0474

Pout
Pout
Pout
Pout
Pout
Pout
Pout

ABSOLUTE ZERO DEGREES
UK

0475

ABSOLUTE ZERO DEGREES
UK

0476

ABSOLUTE ZERO DEGREES
UK
0477

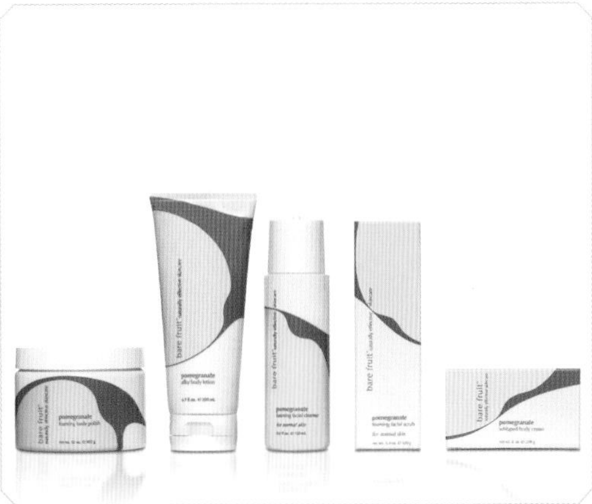

TURNER DUCKWORTH
USA
0478

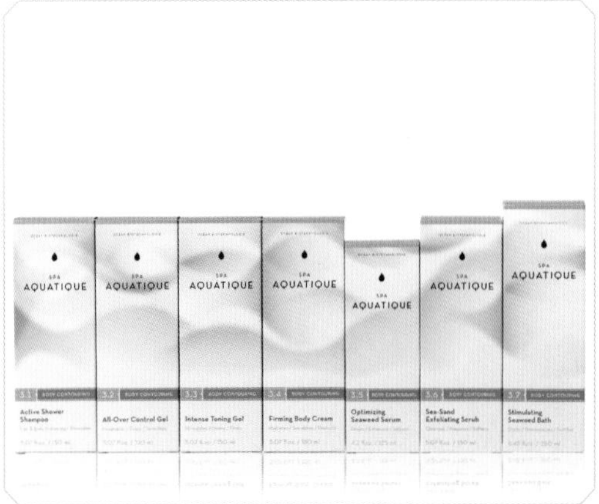

ARCHRIVAL
USA
0479

ALEXANDER ISLEY INC.
USA
0480

ALEXANDER ISLEY INC.
USA
0481

ARCHRIVAL
USA
0482

HORNALL ANDERSON DESIGN
USA
0483

HORNALL ANDERSON DESIGN
USA
0484

HORNALL ANDERSON DESIGN
USA
0485

BOY BASTIAENS
THE NETHERLANDS
0486

ARCHRIVAL
USA
0487

ARCHRIVAL
USA
0488

ARCHRIVAL
USA
0489

ARCHRIVAL
USA
0490

ARCHRIVAL
USA
0491

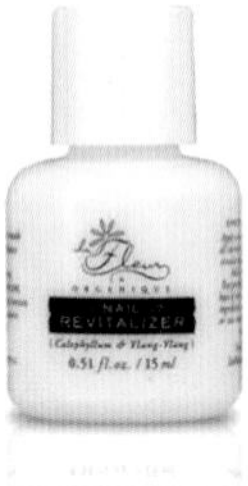

ARCHRIVAL
USA
0492

ARCHRIVAL
USA
0493

ARCHRIVAL
USA
0494

la Fleur
ORGANIQUE
RELAXING
MASSAGE OIL
{ Bergamot & Bitter Orange }
GENTLE TOUCH
BODY CARE
MOISTURIZES AND
SOFTENS
1.69 fl.oz. / 50 ml

la Fleur
ORGANIQUE
EXFOLIATING
SCRUB
{ Lavender & Almond }
GENTLE TOUCH
BODY CARE
MOISTURIZES AND
SOFTENS
5.07 fl.oz. / 150 ml

la Fleur
ORGANIQUE
MOISTURIZING
MILK
{ Lavender & Aloe Vera }
GENTLE TOUCH
BODY CARE
MOISTURIZES AND
SOFTENS
5.07 fl.oz. / 150 ml

la Fleur
ORGANIQUE
FIRMING CREAM
{ Amino Acids, Horse Chestnut, Tea, Coffee, Centella & Grapefruit }
Organic skincare made in Provence, France
GENTLE TOUCH
BODY CARE
MOISTURIZES AND
SOFTENS
5.07 fl.oz. / 150 ml

ARCHRIVAL
USA

0495

FUNNEL : ERIC KASS : UTILITARIAN + COMMERCIAL + FINE : ART
USA

0496

TEA FORTÉ
USA
0497

TEA FORTÉ
USA
0498

TEA FORTÉ
USA
0499

TEA FORTÉ
USA
0500

TURNER DUCKWORTH
USA

0501

TURNER DUCKWORTH
USA

0502

TURNER DUCKWORTH
USA

0503

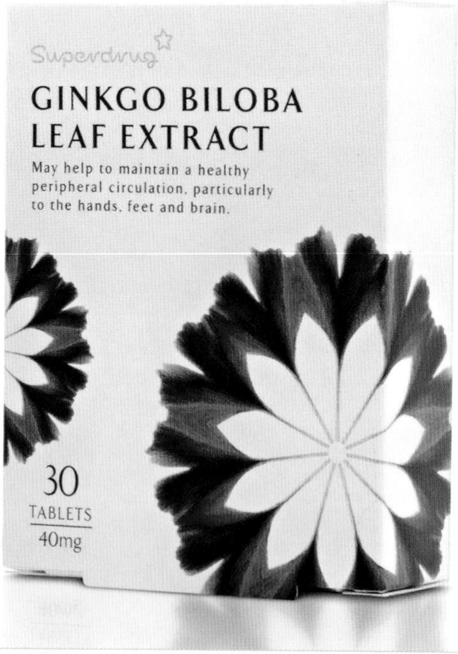

TURNER DUCKWORTH
USA

0504

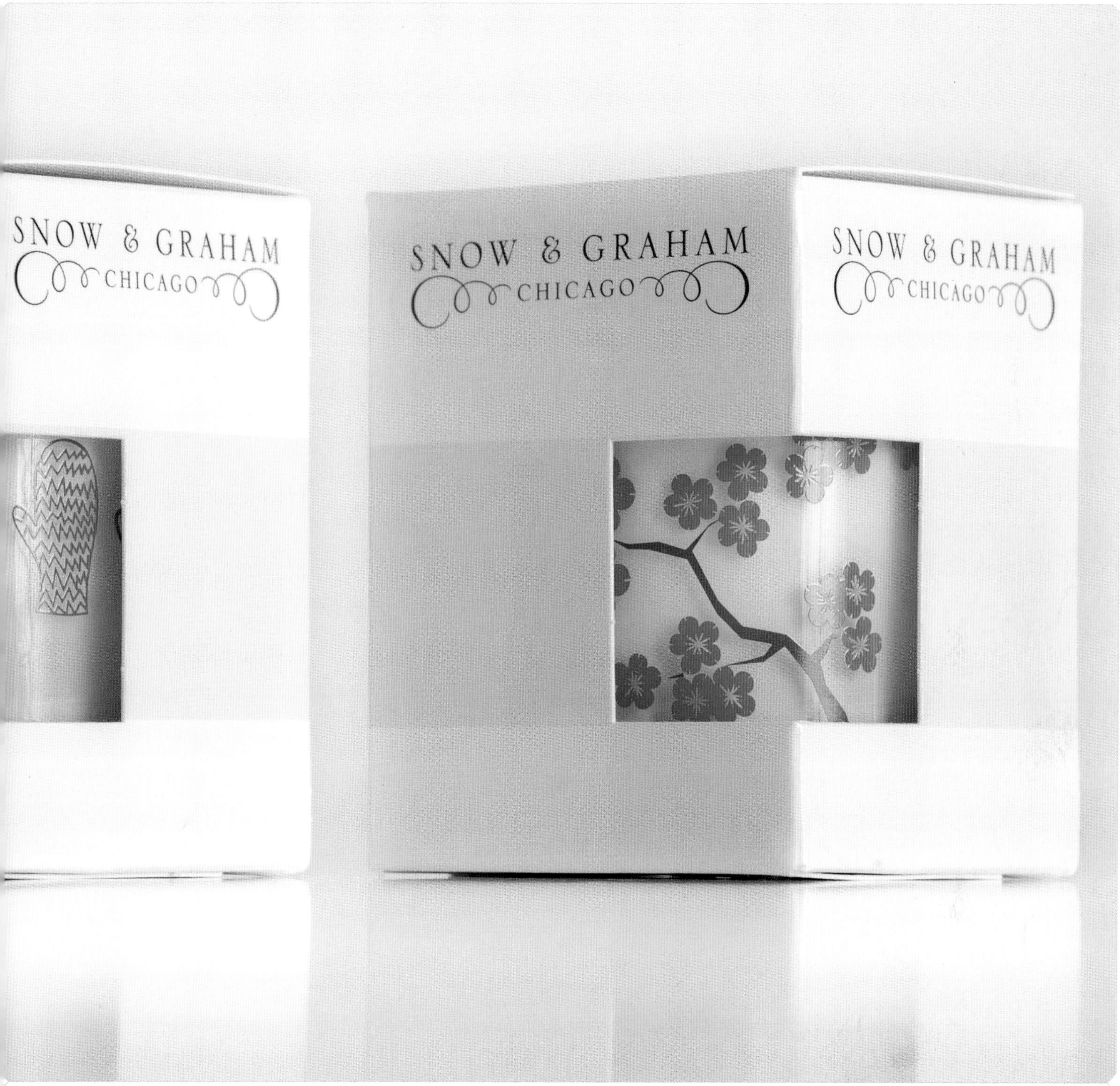

CRASH
USA

0505

CECI NEW YORK
USA

0506

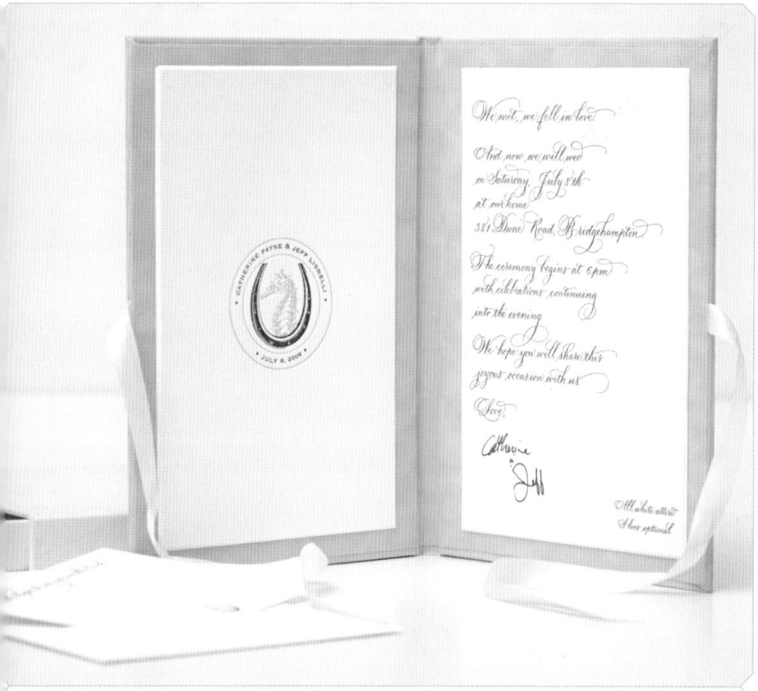
CATHERINE PAYNE & JEFF LIGNELLI
JULY 8, 2006
We met, we fell in love.
And now, we will wed
on Saturday, July 8th
at our home
381 Dune Road, Bridgehampton
The ceremony begins at 6pm
with celebrations continuing
into the evening
We hope you will share this
joyous occasion with us
Love,
Catherine
&
Jeff
All white attire
Shoes optional

CECI NEW YORK
USA

0507

CECI NEW YORK
USA

0508

701 KANUGA DRIVE
WEST PALM BEACH, FLORIDA 33401
Mr. & Mrs. Alexander Brookstep
APARTMENT 14G
32123 MAYCOMB ROAD
WESTING, LOUISIANA 33401
Save the Date

CECI NEW YORK
USA

0509

Save the Date
To celebrate the marriage of

CECI NEW YORK
USA

0510

BUCHANAN DESIGN
USA
0511

BUCHANAN DESIGN
USA
0512

JOSHUA BLAYLOCK
USA
0513

CECI NEW YORK
USA
0514

Our wedding is
fast approaching!
Please
save the date
June 30 2007
to celebrate the
marriage of
Neha Patel
Нейха Пател
and
Boris
Markovich
345 East 73rd Street
Apartment 10 F
New York, New York 10021
DR. ALYSSA BRUNET
DR. SAM GAMOKA
3160 OCEAN AVENUE
APARTMENT B2
BROOKLYN, NEW YORK
11235

CECI NEW YORK
USA

0515

the story of
nikki & nick
by mary olivieri
the story of
nikki & nick
Larry and Jand Niehus
618 Coniston Circle
Sergeant Bluff IA 510547
nikki & nick

CALIENTE CREATIVE
USA

0516

CECI NEW YORK
USA
0517

BUNGALOW CREATIVE
USA
0518

BUNGALOW CREATIVE
USA
0519

EPOS, INC.
USA
0520

THE COLLECTIVE DESIGN CONSULTANTS
AUSTRALIA
0521

THE COLLECTIVE DESIGN CONSULTANTS
AUSTRALIA
0522

THE COLLECTIVE DESIGN CONSULTANTS
AUSTRALIA
0523

THE COLLECTIVE DESIGN CONSULTANTS
AUSTRALIA
0524

MELBOURNE LOUNGE
Cabernet Merlot
MELBOURNE LOUNGE
Rosé
MELBOURNE LOUNGE
Chardonnay Pinot Grigio

DESIGN BRIDGE
UNITED USA

0525

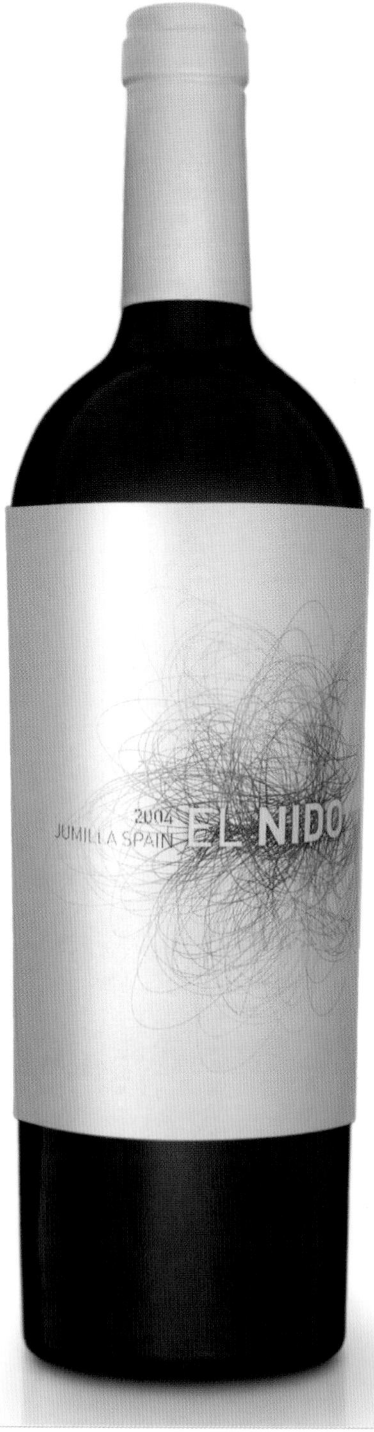
2004
JUMILLA SPAIN
EL NIDO

GO BIG
USA

0526

TURNER DUCKWORTH
USA

0527

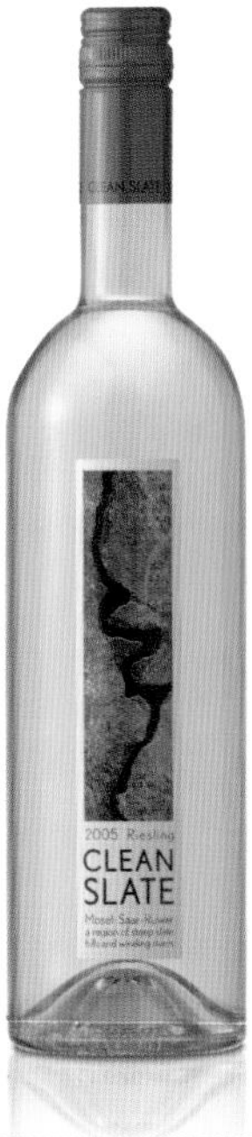

TURNER DUCKWORTH
USA

0528

GO BIG
USA

0529

THE COLLECTIVE DESIGN CONSULTANTS
AUSTRALIA

0530

KENDALL ROSS
USA
0531

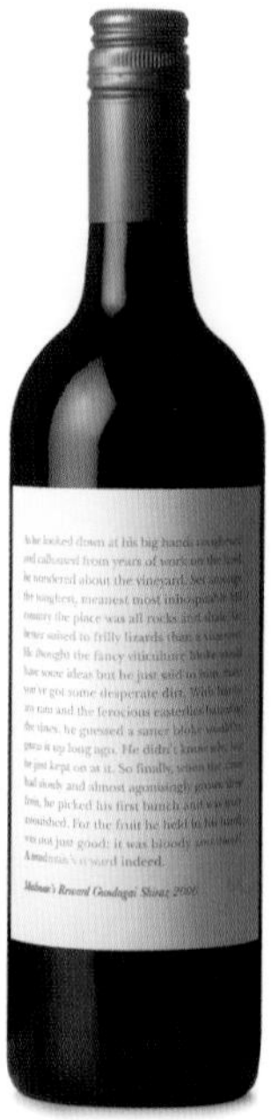

THE COLLECTIVE DESIGN CONSULTANTS
AUSTRALIA
0532

TURNER DUCKWORTH
USA
0533

TURNER DUCKWORTH
USA
0534

antipodes
SPARKLING
To be at your table today this water has been brought to the surface from the deepest water aquifer in New Zealand. It has spent decades under enormous pressure in vast underground canyons more than 200 metres below the surface. This pressure from within the aquifer creates a natural filtration process that has led to antipodes being scientifically categorised as the deepest, highest quality artesian water in New Zealand. It has then been bottled at source providing a purity, clarity and taste that can only be found deep down at the end of the earth. Gently carbonated with the finest bead, antipodes is the perfect partner for fine foods.
DRINK CHILLED. DRINK OFTEN. LIVE WELL.
1000ml
Antipodes Water Company, 121 Customs St West, Auckland
New Zealand Artesian Water www.antipodes.co.nz

ANTIPODES WATER CO.
NEW ZEALAND

0535

DIESEL
FUEL FOR LIFE
75 ML
USE WITH CAUTION

L'OREAL USA — DIESEL FRAGRANCES
USA

0536

IRVING
UK

0537

IRVING
UK

0538

GRIP
USA

0539

MORLA DESIGN
USA

0540

HELENA SEO DESIGN
USA
0541

HELENA SEO DESIGN
USA
0542

HELENA SEO DESIGN
USA
0543

HELENA SEO DESIGN
USA
0544

HELENA SEO DESIGN
USA

0545

DIESEL
FUEL FOR LIFE
75 ML
USE WITH CAUTION

L'OREAL USA — DIESEL FRAGRANCES
USA

0546

HELENA SEO DESIGN
USA

0547

HELENA SEO DESIGN
USA

0548

HELENA SEO DESIGN
USA

0549

HELENA SEO DESIGN
USA

0550

AGED 15 AÑOS
AGED 15 AÑOS
BOTTLED IN:
2005
BOTTLED BY:
BOTTLE NUMBER:
no.8471

P&W
UK

0551

LOSELEY
DAIRY ICE CREAM
DOUBLE BELGIAN
CHOCOLATE
WITH REAL JERSEY CREAM

P&W
UK

0552

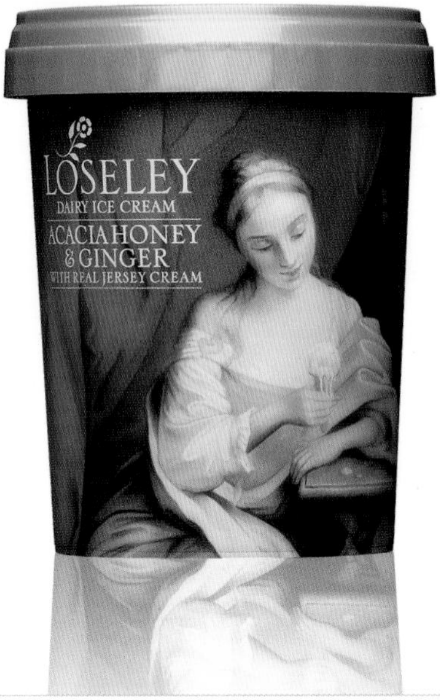
LOSELEY
DAIRY ICE CREAM
ACACIA HONEY
& GINGER
WITH REAL JERSEY CREAM

P&W
UK

0553

LOSELEY
DAIRY ICE CREAM
STRAWBERRIES
& CREAM
WITH REAL JERSEY CREAM

P&W
UK

0554

AGED 10 AÑOS
Pure Cask-Aged Mexican Rum
CASK FILLED 1994 BOTTLED 2005
RON VERACRUZANO
AGED 10 AÑOS
AÑEJO
Produced & bottled in Mexico
50cl.e 40%Vol.
PRODUCED & BOTTLED BY: LICORES VERACRUZ SA DE CV CORDOBA (VE), MEXICO
AGED 15 AÑOS
Pure Cask-Aged Mexican Rum
CASK FILLED 1989 BOTTLED 2005
RON VERACRUZANO
AGED 15 AÑOS
AÑEJO ESPECIAL
Produced & bottled in Mexico
50cl.e 40%Vol.
PRODUCED & BOTTLED BY: LICORES VERACRUZ SA DE CV CORDOBA (VE), MEXICO
BOTTLE NUMBER: No.8471
BOTTLED IN: 2005
NO.8471
Pure Cask-Aged Mexican Rum
CASK FILLED 1984 BOTTLED 2005
RON VERACRUZANO
AGED 20 AÑOS
AÑEJO RESERVA ESPECIAL
Produced & bottled in Mexico
50cl.e 43%Vol.
PRODUCED & BOTTLED BY: LICORES VERACRUZ SA DE CV CORDOBA (VE), MEXICO

P&W
UK

0555

ROOM №
6
LOUNGE
CARDAMOM & TONKA BEAN
NET 26 OZ
PADDYWAX
100+ HOURS

PRINCIPLE
USA

0556

MILLER CREATIVE
USA
0557

MILLER CREATIVE
USA
0558

PEARLFISHER
USA
0559

DAVID CARTER DESIGN ASSOCIATES
USA
0560

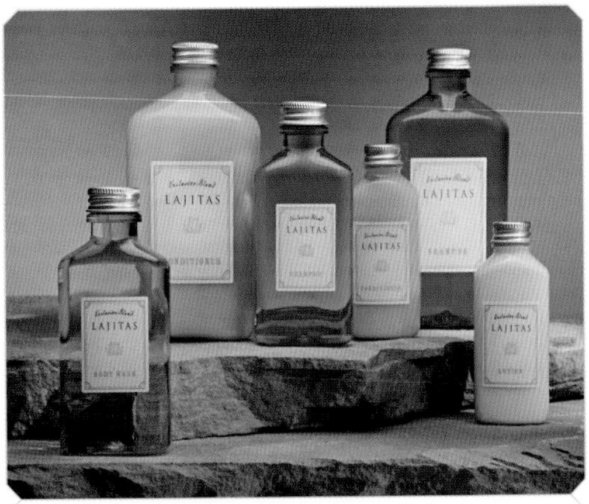

DAVID CARTER DESIGN ASSOCIATES
USA
0561

UNIVERSITY OF NEBRASKA AT KEARNEY
USA
0562

MORLA
USA
0563

PHILIPPE BECKER DESIGN
USA
0564

THE O GROUP
USA
0565

SONSOLES LLORENS
SPAIN
0566

SONSOLES LLORENS
SPAIN
0567

SONSOLES LLORENS
SPAIN
0568

SONSOLES LLORENS
SPAIN
0569

ELEGANT & ENGLISH
HONEY & ALMOND
An infusion of two
Elegant flavours in an
English all-butter biscuit
CHOCOLATE & RASPBERRY
GINGER & LEMON
ROSEHIP & BLACKCURRANT
COCONUT & LIME
VANILLA & CREAM

IRVING
UK

0570

thinkThin
CHUNKY PEANUT BUTTER
thinkGreen
CRANBERRY APPLE
superfood
nutrition bar
for optimum
vibrance
thinkOrganic
CHERRY NUT
raw fruit
vitality

ADDIS CRESON
USA

0571

TOKY BRANDING + DESIGN
USA
0572

TOKY BRANDING + DESIGN
USA
0573

TOKY BRANDING + DESIGN
USA
0574

TURNER DUCKWORTH
USA
0575

TURNER DUCKWORTH
USA
0576

TURNER DUCKWORTH
USA
0577

BAKKEN CREATIVE CO.
USA
0578

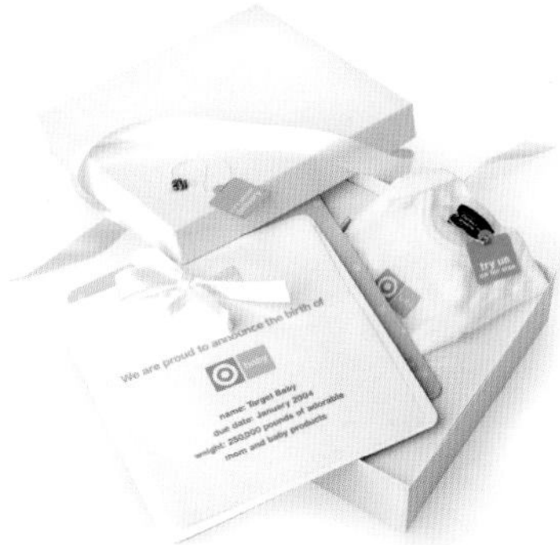

CATALYST STUDIOS
USA
0579

EVENSON DESIGN GROUP
USA
0580

BISSINGER'S
THE CHOICE OF KINGS AND CONNOISSEURS
Bissinger's
SUGAR FREE
MOCHA
GOURMET COCOA BLEND
GOURMET BLENDS
NET WT. 8 OZ.
[227 GM.]
EST. IN EUROPE 1668

TOKY BRANDING + DESIGN
USA

0581

TOKY BRANDING + DESIGN
USA
0582

TOKY BRANDING + DESIGN
USA
0583

TOKY BRANDING + DESIGN
USA
0584

TOKY BRANDING + DESIGN
USA
0585

TOKY BRANDING + DESIGN
USA
0586

TOKY BRANDING + DESIGN
USA
0587

TOKY BRANDING + DESIGN
USA
0588

TOKY BRANDING + DESIGN
USA
0589

TOKY BRANDING + DESIGN
USA
0590

TOKY BRANDING + DESIGN
USA
0591

TOKY BRANDING + DESIGN
USA
0592

TOKY BRANDING + DESIGN
USA
0593

TOKY BRANDING + DESIGN
USA
0594

GANACHÈ TRUFFLES
THE CHOICE OF KINGS AND CONNOISSEURS
Bissinger's
Handcrafted Chocolatier
8
GANACHÈ
TRUFFLES
Blueberry, Double Chocolate, Cranberry, Crème de Menthe, Raspberry, Coconut, Blackberry and Irish Crème
GANACHÈ LIQUEUR TRUFFLES
THE CHOICE OF KINGS AND CONNOISSEURS
Bissinger's
Handcrafted Chocolatier
8
GANACHÈ LIQUEUR
TRUFFLES
Amaretto, Rum, Brandy and Grand Marnier

TOKY BRANDING + DESIGN
USA

0595

FRESH SEAL
PEPPERIDGE FARM
AUTHENTIC ARTISAN
BREAD
TASTES FRESH FROM THE BAKERY
10
MIN
ONE SOURDOUGH BOULE
NET WT. 1LB (16oz.)

FRESH SEAL
PEPPERIDGE FARM
AUTHENTIC ARTISAN
BREAD
TASTES FRESH FROM THE BAKERY
10
MIN
ONE ROSEMARY OLIVE OIL BOULE
NET WT. 1LB (16oz.)

WALLACE CHURCH
USA

0596

WALLACE CHURCH
USA
0597

MICHAEL OSBORNE DESIGN
USA
0598

MICHAEL OSBORNE DESIGN
USA
0599

BAKKEN CREATIVE CO.
USA
0600

TOKY BRANDING + DESIGN
USA
0601

SONSOLES LLORENS
SPAIN
0602

SONSOLES LLORENS
SPAIN
0603

SONSOLES LLORENS
SPAIN
0604

SONSOLES LLORENS
SPAIN
0605

BUCHANAN DESIGN
USA

0606

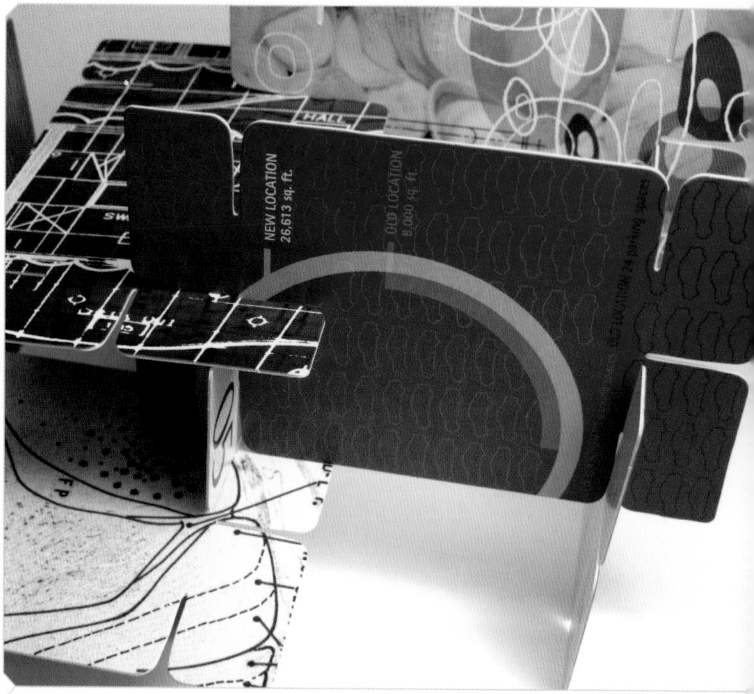

BUCHANAN DESIGN
USA

0607

SONSOLES LLORENS
SPAIN

0608

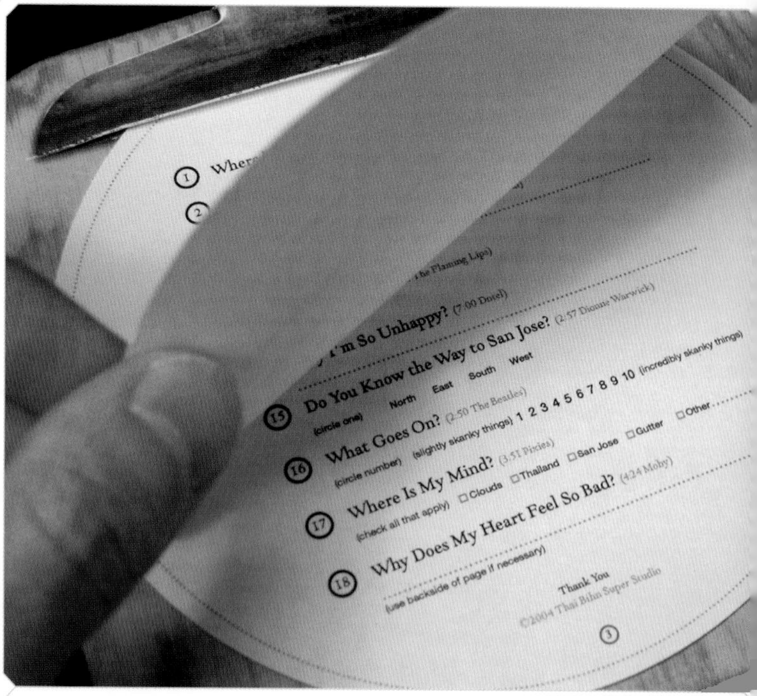

STUDIOBENBEN
USA

0609

WILD BLACKBERRY
the journey of the bee
CLOVER
natural beeswax blend
NET 14 OZ
80 HOURS
FREESIA
natural beeswax blend
NET 14 OZ
80 HOURS

PRINCIPLE
USA

0610

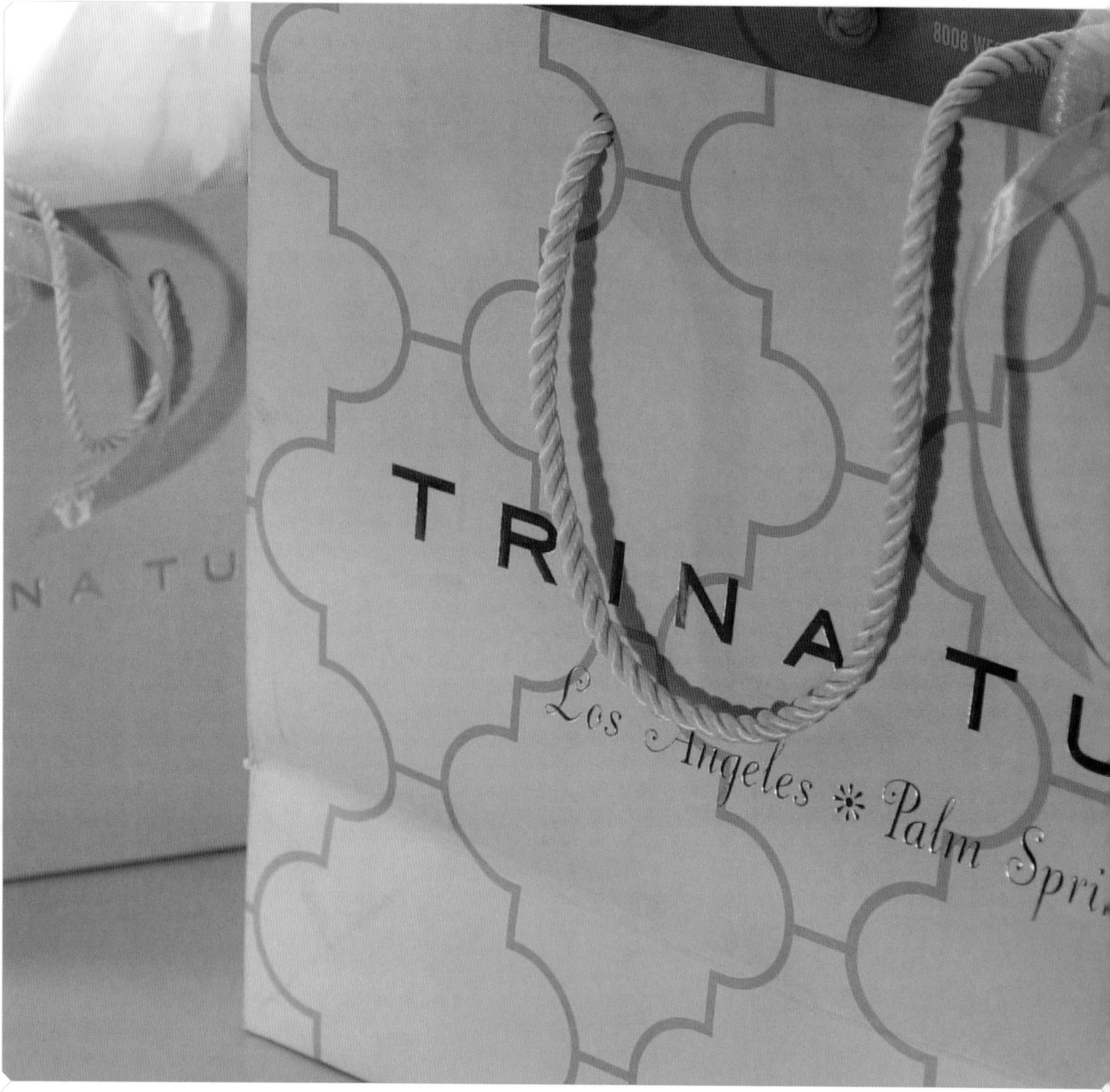
TRINA TU
Los Angeles * Palm Spri

BLUE LOUNGE DESIGN
USA

0611

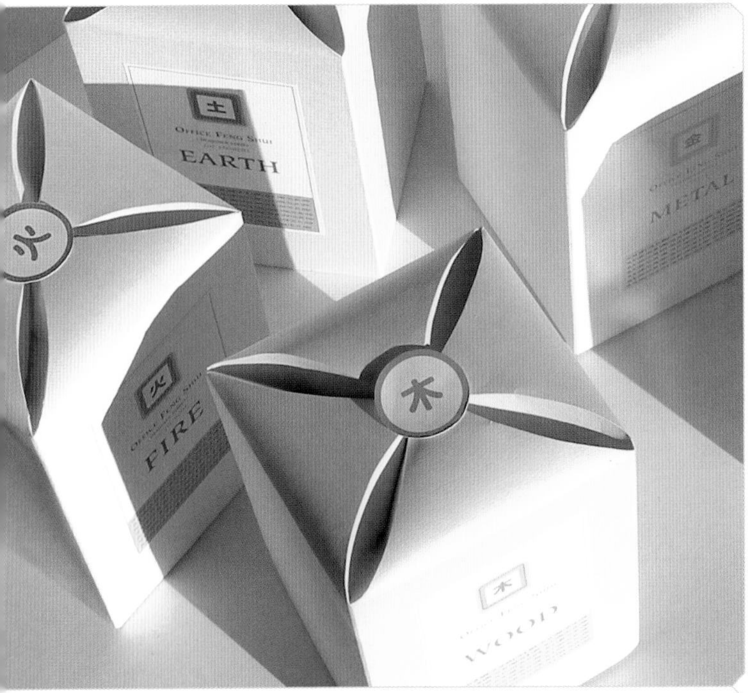

KAYU TAI
USA

0612

MICHAEL OSBORNE DESIGN
USA

0613

PRINCIPLE
USA

0614

IRVING
UK

0615

PHILIPPE BECKER DESIGN
USA
0616

HOMART
USA
0617

TOKY BRANDING + DESIGN
USA
0618

SONSOLES LLORENS
SPAIN
0619

bright ideas
oops
virgin atlantic
forty winks
virgin atlantic
virgin atlantic
pardon?
clean
polish
virgin atlantic
virgin atlantic

TURNER DUCKWORTH
USA

0620

Quinta
da Casa Boa
2005
vinho tinto
Vinho regional Estremadura
Casa Boa
2005
vinho tinto
Vinho regional Estremadura

JULIA JURIGA-LAMUT
AUSTRIA

0621

JULIA JURIGA-LAMUT
AUSTRIA
0622

DESIGN AHEAD
GERMANY
0623

DESIGN AHEAD
GERMANY
0624

MICHAEL OSBORNE DESIGN
USA
0625

SPARK STUDIO
AUSTRALIA
0626

SPARK STUDIO
AUSTRIA
0627

OCTAVO DESIGN
AUSTRALIA
0628

OCTAVO DESIGN
AUSTRALIA
0629

OCTAVO DESIGN
AUSTRALIA
0630

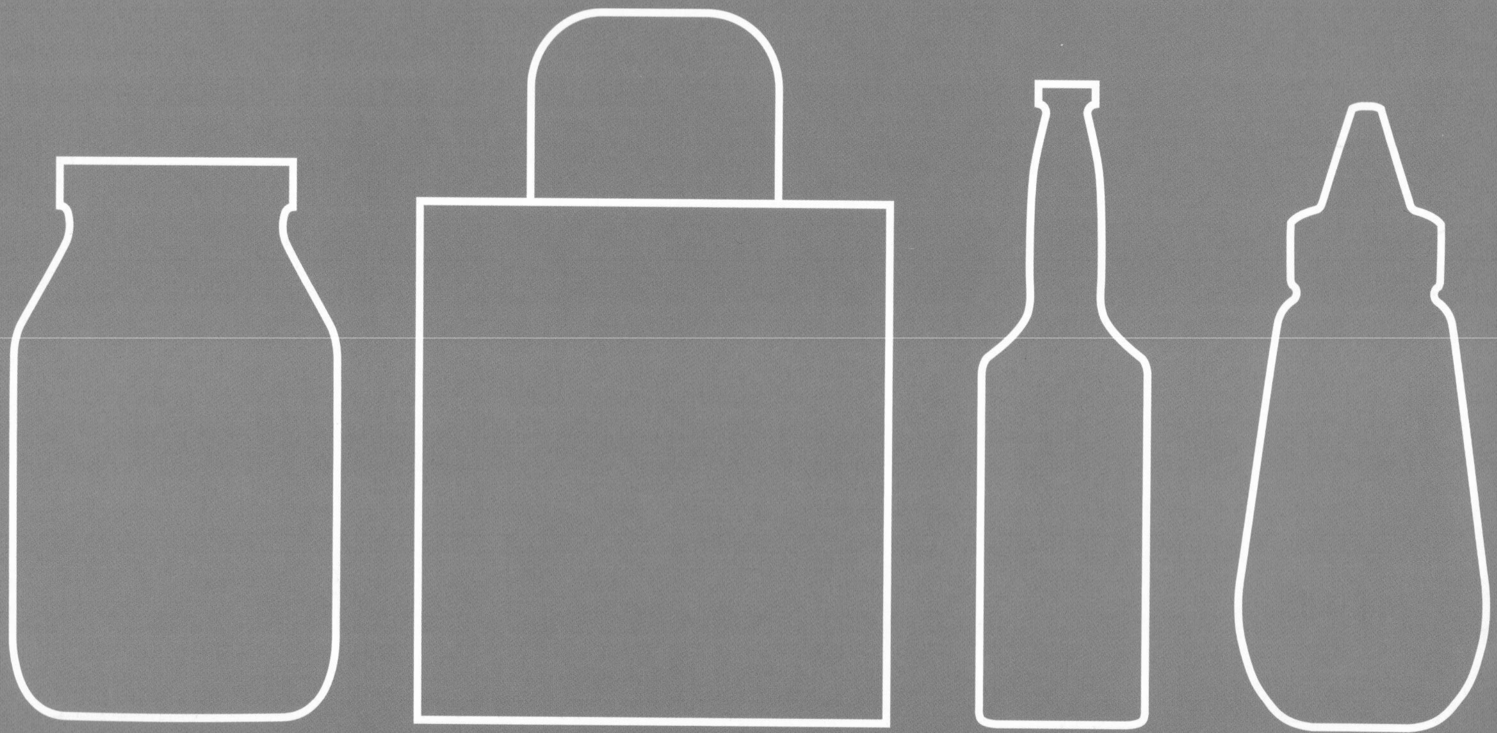

wow!

PRACTICAL
MASS-APPEAL
STRAIGHT-FORWARD
PROFICIENT
SMART
PLEASANT
EFFECTIVE
ATTRACTIVE
ASTUTE

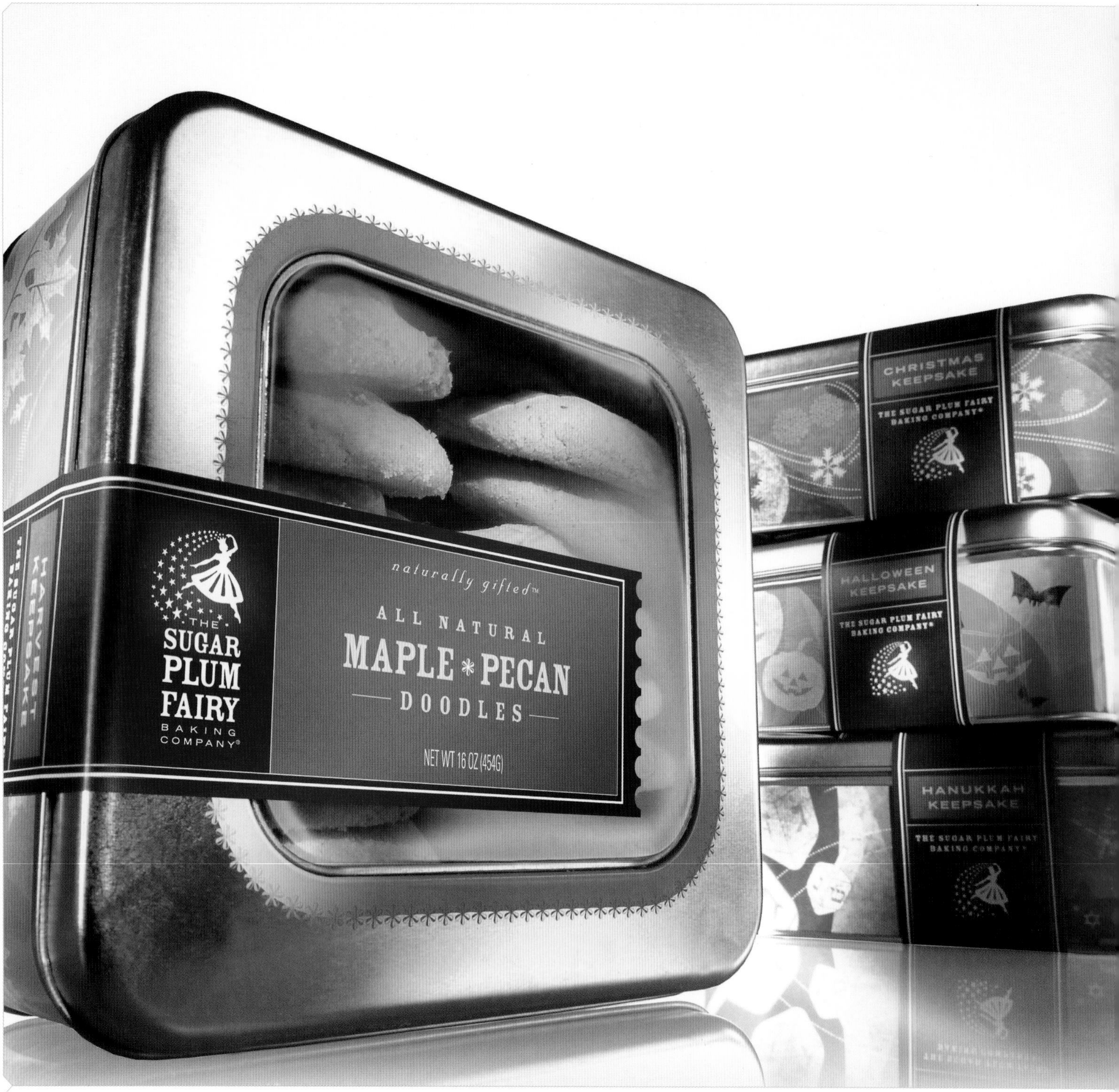
THE SUGAR PLUM FAIRY BAKING COMPANY®
naturally gifted™
ALL NATURAL
MAPLE * PECAN
DOODLES
NET WT 16 OZ (454G)
HARVEST KEEPSAKE
CHRISTMAS KEEPSAKE
THE SUGAR PLUM FAIRY BAKING COMPANY®
HALLOWEEN KEEPSAKE
THE SUGAR PLUM FAIRY BAKING COMPANY®
HANUKKAH KEEPSAKE
THE SUGAR PLUM FAIRY BAKING COMPANY®

CRAVE INC.
USA

0631

CRAVE INC.
USA
0632

CRAVE INC.
USA
0633

CRAVE INC.
USA
0634

CRAVE INC.
USA
0635

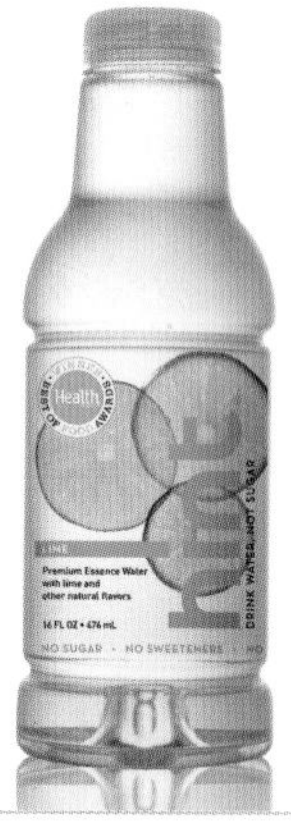

BLUE LOUNGE DESIGN
USA
0636

BLUE LOUNGE DESIGN
USA
0637

BLUE LOUNGE DESIGN
USA
0638

BLUE LOUNGE DESIGN
USA
0639

BLUE LOUNGE DESIGN
USA
0640

BLUE LOUNGE DESIGN
USA
0641

BLUE LOUNGE DESIGN
USA
0642

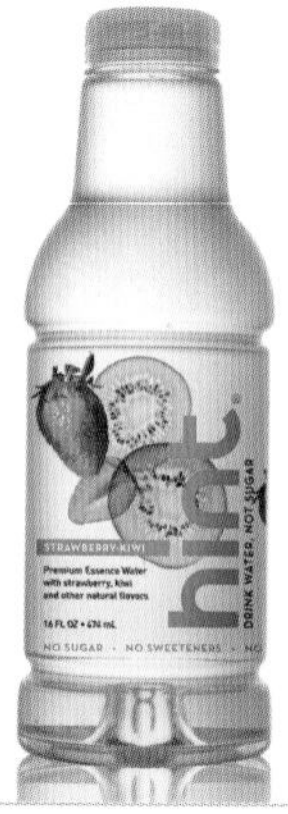

BLUE LOUNGE DESIGN
USA
0643

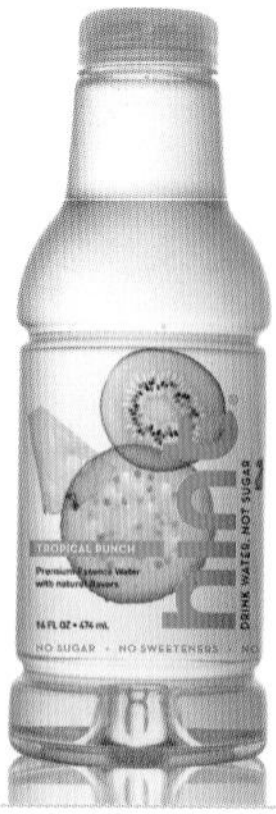

BLUE LOUNGE DESIGN
USA
0644

FROM THE HINTERLAND
IMMEDIATE SEATING IN THE CITRUS SECTIONS
Botanically speaking, this tangy variant of the mandarin orange, doesn't even exist—yet, you hold its essence in a bottle. And, why are they called "tangerines" anyway? Long revered in Asia, both for their refreshing flavors and healing powers, citrus fruit were believed an effective antidote to scorpion bites. Beware.
TANGERINE
Premium Drinking Water
No Sugar • All Natural
15 FL OZ • 444 mL
hint™
DRINK WATER, NOT SUGAR
ANSWER:
Laurel and Hardy, of course.
PEAR
Premium Drinking Water
No Sugar • All Natural
hint™
DRINK WATER, NOT SUGAR
FROM THE HINTERLAND
THE APP-SOLUTE TRUTH
They say that, "An apple a day keeps the doctor away." Does that mean that eating one is like taking a pill? Get it? Aaa-pille? Which brings us to our next question: Did an apple really fall on Isaac Newton's head?
APPLE
Premium Drinking Water
No Sugar • All Natural
15 FL OZ • 444 mL
hint kids™
DRINK WATER, NOT SUGAR

BLUE LOUNGE DESIGN
USA

0645

P&W
UK
0646

P&W
UK
0647

BUNGALOW CREATIVE
USA
0648

P&W
UK
0649

PHILIPPE BECKER DESIGN
USA

0650

P&W
UK

0651

P&W
UK

0652

FUNNEL: ERIC KASS: UTILITARIAN + COMMERCIAL + FINE: ART
USA

0653

O! ADVERTISING & DESIGN
ICELAND
0654

O! ADVERTISING & DESIGN
ICELAND
0655

O! ADVERTISING & DESIGN
ICELAND
0656

O! ADVERTISING & DESIGN
ICELAND
0657

WALLACE CHURCH, INC.
USA
0658

WALLACE CHURCH, INC.
USA
0659

CREATIVE COMMUNE / URBAN ACCENTS
USA
0660

CREATIVE COMMUNE / URBAN ACCENTS
USA
0661

CREATIVE COMMUNE / URBAN ACCENTS
USA
0662

IRVING
UK

0663

ARTISAN
ASHBOURNE
BAKED BY HAND
ENGLAND
BISCUITS
TWO BY TWO
Organic
Strawberry Biscuits
8 little Lion and Mouse shaped all butter
biscuits with real strawberry juice
ARTISAN
ASHBOURNE
BAKED BY HAND
ENGLAND
BISCUITS
TWO BY TWO
Organic
Toffee Biscuits
ARTISAN
ASHBOURNE
BAKED BY HAND
ENGLAND
BISCUITS
TWO BY TWO
Organic
Vanilla Biscuits
8 little Owl and Pussycat shaped all butter
biscuits made with Madagascan Vanilla extract
£5

SWAK
the essential handbook of
HOLIDAY GREETINGS
a Merry Mistletoe Kit from The Jones Group
DEC
SEALED WITH A
SWAK

THE JONES GROUP
USA

0664

DELTA ENTERTAINMENT CORPORATION
USA

0665

DELTA ENTERTAINMENT CORPORATION
USA

0666

DELTA ENTERTAINMENT CORPORATION
USA

0667

DELTA ENTERTAINMENT CORPORATION
USA

0668

TOKY BRANDING + DESIGN
USA
0669

SONSOLES LLORENS
SPAIN
0670

CRAVE, INC.
USA
0671

ROME & GOLD CREATIVE
USA
0672

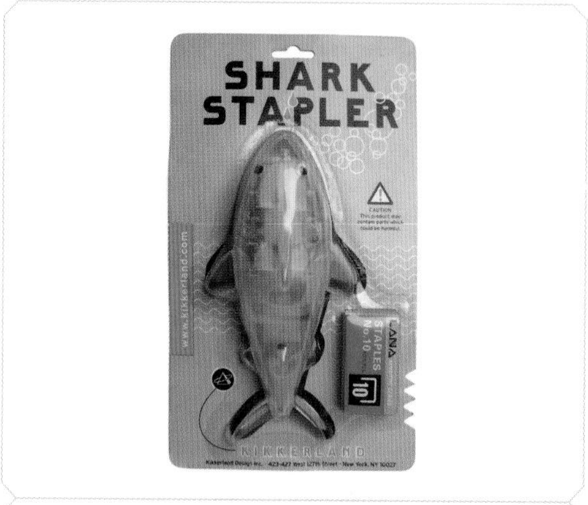

212-BIG-BOLT
USA
0673

OKOLITA M
USA
0674

DESIGN FORCE, INC.
USA
0675

BUNGALOW CREATIVE
USA
0676

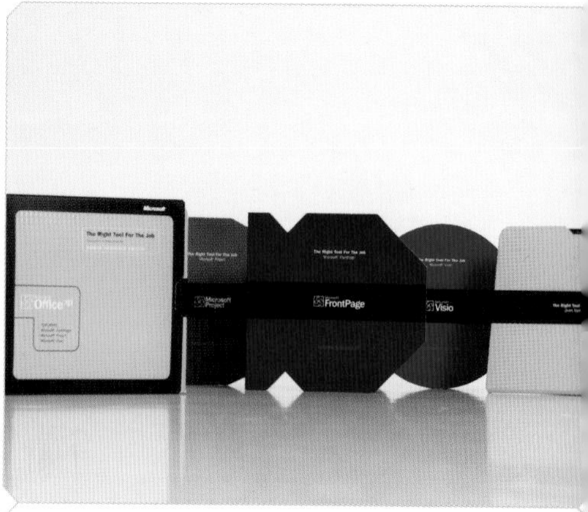

FULLBLASTINC.COM
USA
0677

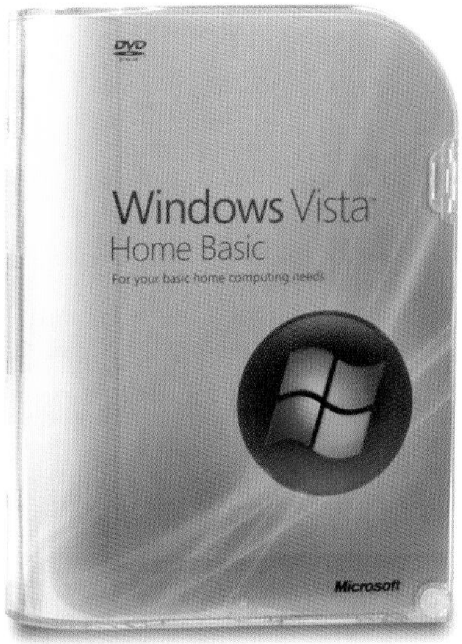

HORNALL ANDERSON DESIGN
USA
0678

HORNALL ANDERSON DESIGN
USA
0679

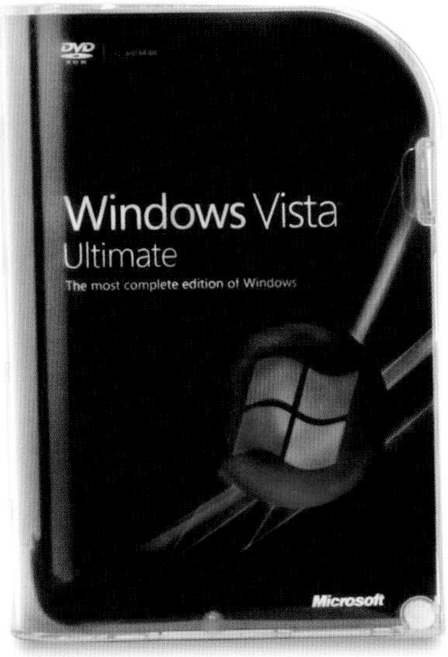

HORNALL ANDERSON DESIGN
USA
0680

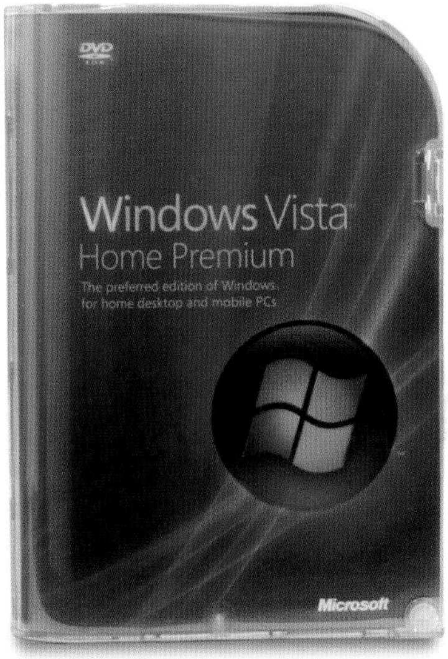

HORNALL ANDERSON DESIGN
USA
0681

Waitrose
Lightweight
Cat litter
Absorbent
Helps to Control Odours
℮ 10 Litre

TURNER DUCKWORTH
USA

0682

TURNER DUCKWORTH
USA
0683

TURNER DUCKWORTH
USA
0684

HORNALL ANDERSON DESIGN
USA
0685

TURNER DUCKWORTH
USA
0686

TURNER DUCKWORTH
USA
0687

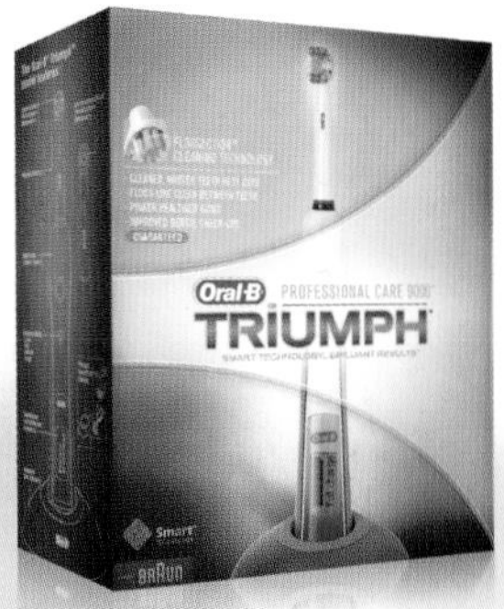

WALLACE CHURCH, INC.
USA
0688

SPARK STUDIO
AUSTRALIA
0689

TURNER DUCKWORTH
USA
0690

TURNER DUCKWORTH
USA
0691

TURNER DUCKWORTH
USA
0692

P&W
UK
0693

NIEDERMEIER DESIGN
USA
0694

NIEDERMEIER DESIGN
USA
0695

NIEDERMEIER DESIGN
USA
0696

NIEDERMEIER DESIGN
USA
0697

P&W
UK
0698

MICHAEL OSBORNE DESIGN
USA
0699

PHILIPPE BECKER DESIGN
USA
0700

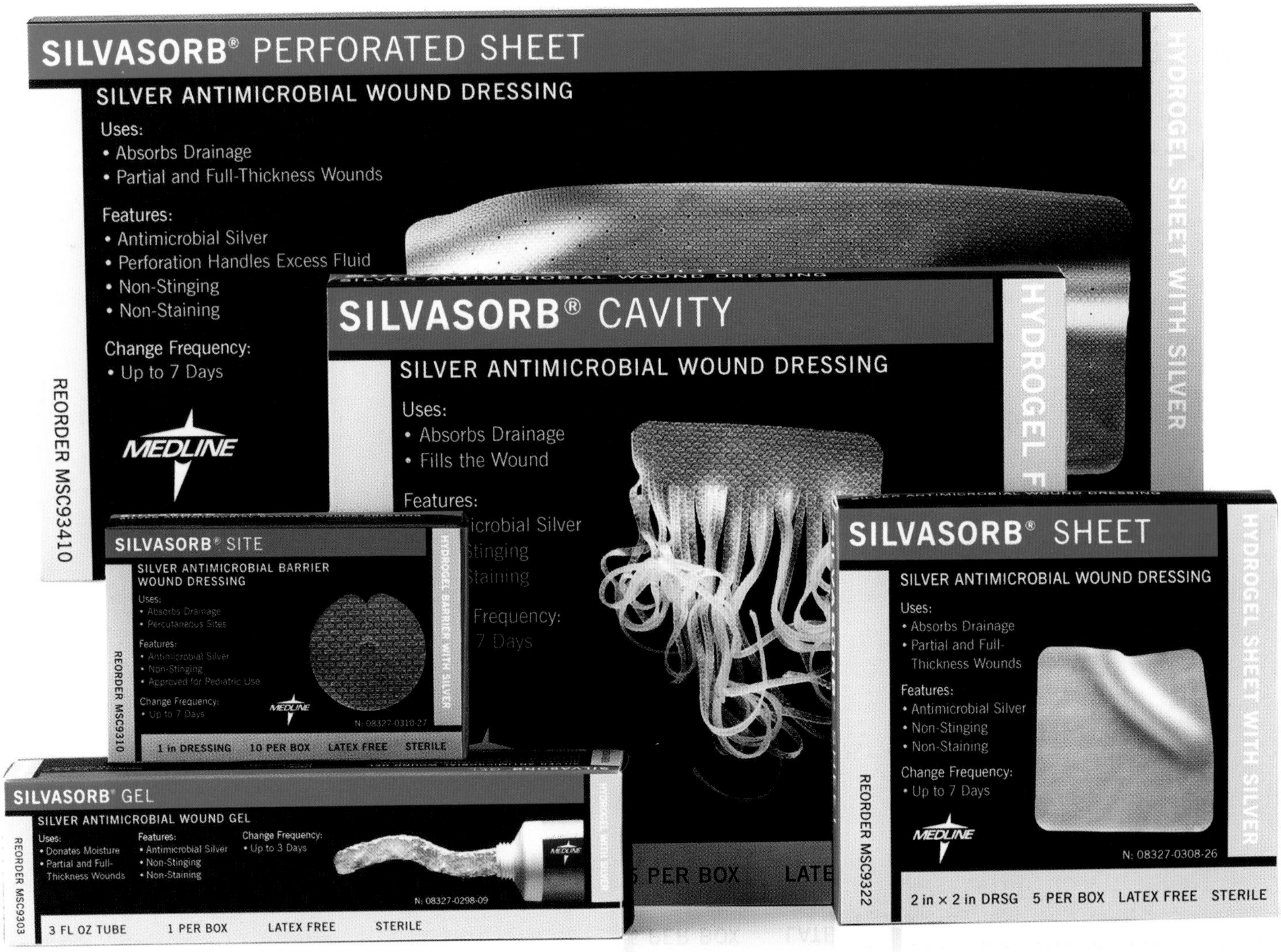
SILVASORB® PERFORATED SHEET
SILVER ANTIMICROBIAL WOUND DRESSING
Uses:
• Absorbs Drainage
• Partial and Full-Thickness Wounds
Features:
• Antimicrobial Silver
• Perforation Handles Excess Fluid
• Non-Stinging
• Non-Staining
Change Frequency:
• Up to 7 Days
MEDLINE
REORDER MSC93410
HYDROGEL SHEET WITH SILVER
SILVASORB® CAVITY
SILVER ANTIMICROBIAL WOUND DRESSING
Uses:
• Absorbs Drainage
• Fills the Wound
Features:
SILVASORB® SITE
SILVER ANTIMICROBIAL BARRIER WOUND DRESSING
REORDER MSC9310
HYDROGEL BARRIER WITH SILVER
1 in DRESSING 10 PER BOX LATEX FREE STERILE
SILVASORB® GEL
SILVER ANTIMICROBIAL WOUND GEL
Uses:
• Donates Moisture
• Partial and Full-Thickness Wounds
Features:
• Antimicrobial Silver
• Non-Stinging
• Non-Staining
Change Frequency:
• Up to 3 Days
REORDER MSC9303
HYDROGEL WITH SILVER
N: 08327-0298-09
3 FL OZ TUBE 1 PER BOX LATEX FREE STERILE
SILVASORB® SHEET
SILVER ANTIMICROBIAL WOUND DRESSING
Uses:
• Absorbs Drainage
• Partial and Full-Thickness Wounds
Features:
• Antimicrobial Silver
• Non-Stinging
• Non-Staining
Change Frequency:
• Up to 7 Days
MEDLINE
REORDER MSC9322
HYDROGEL SHEET WITH SILVER
N: 08327-0308-26
2 in × 2 in DRSG 5 PER BOX LATEX FREE STERILE

MILTON GLASER, INC.
USA

0701

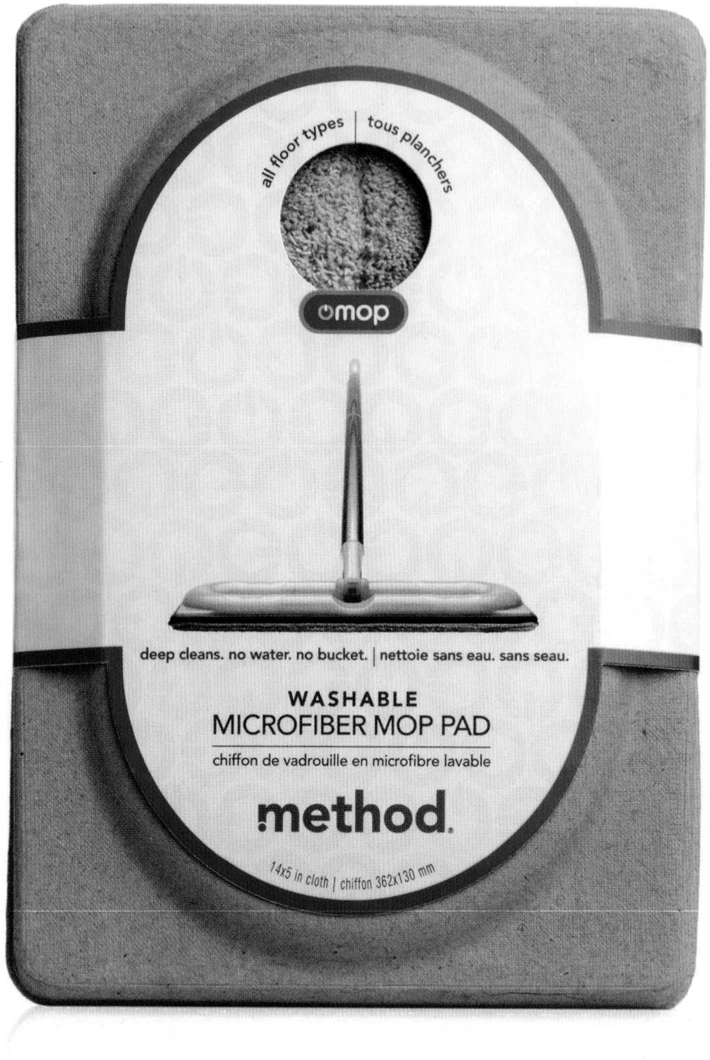
all floor types | tous planchers
mop
deep cleans. no water. no bucket. | nettoie sans eau. sans seau.
WASHABLE
MICROFIBER MOP PAD
chiffon de vadrouille en microfibre lavable
method
14x5 in cloth | chiffon 362x130 mm

METHOD
USA

0702

METHOD
USA
0703

THE JONES GROUP
USA
0704

BLUE LOUNGE DESIGN
USA
0705

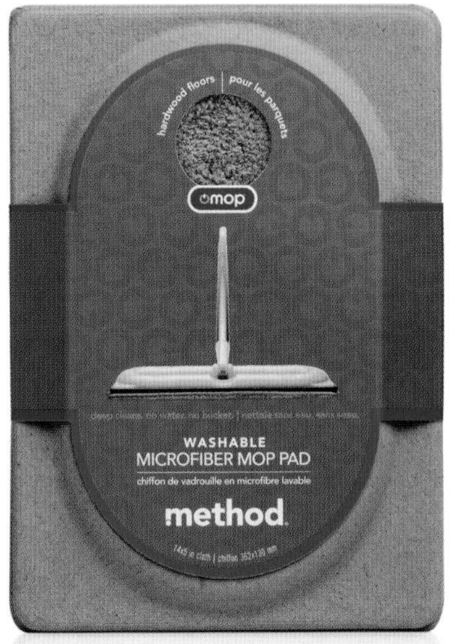

METHOD
USA
0706

LMS DESIGN INC.
USA
0707

MICHAEL OSBORNE DESIGN
USA
0708

KYM ABRAMS DESIGN
USA
0709

MARY HUTCHISON DESIGN LLC
USA
0710

MARY HUTCHISON DESIGN LLC
USA
0711

MARY HUTCHISON DESIGN LLC
USA
0712

CRAVE, INC.
USA
0713

CRAVE, INC.
USA
0714

DESIGN AHEAD
GERMANY
0715

StudioScape
StudioScape lets you change the way you see things... starting with what's directly in front of you. Who can be productive surrounded by clutter? StudioScape helps free your workspace and mind space so you can be more creative, organized and focused. Surrounded by the beauty and simplicity of StudioScape, there's no limit to what you can accomplish!
IDEA GALLERY
Join the O-Scapes Community! E-mail new configurations to MyO-Scapes@oscapes.com and view photos of new ideas at www.O-Scapes.com.
Your idea could go here!
WWW.OSCAPES.COM
UPC HERE
oscapes
StudioScape
THE DESKTOP ACCESSORY WITH BRAINS AND BEAUTY!
Simplify.
Modify.
Beautify.
StudioScape streamlines your desk allowing maximum performance with minimum footage.
So transform your office desk into a fun, personalized studio-oasis with StudioScape!

EVENSON DESIGN GROUP
USA

0716

NO ALCOHOL | SAFE FOR KIDS
Clean Well
Natural
Hand Sanitizing Wipe
Made with
INGENIUM
Kills 99.99% of
germs naturally
NO ALCOHOL | SAFE FOR KIDS
ONE WIPE 6" X 9" (15.2 CM X 22.9 CM)
Clean Well
Natural
Hand Sanitizer
Made with
INGENIUM
Kills 99.99% of
germs naturally
100 sprays
0.5 FL OZ (15 ML)
Clean Well
Natural
Hand Sanitizer
Made with
INGENIUM
Kills 99.99% of
germs naturally
200 sprays
FL OZ (30 ML)

PHILIPPE BECKER DESIGN
USA

0717

28 LIMITED BRAND
GERMANY
0718

CRAVE, INC.
USA
0719

MIRIELLO GRAFICO
USA
0720

WALLACE CHURCH, INC.
USA
0721

CREATIVE COMMUNE / URBAN ACCENTS
USA
0722

CREATIVE COMMUNE / URBAN ACCENTS
USA
0723

CREATIVE COMMUNE / URBAN ACCENTS
USA
0724

CREATIVE COMMUNE / URBAN ACCENTS
USA
0725

CREATIVE COMMUNE / URBAN ACCENTS
USA
0726

CREATIVE COMMUNE / URBAN ACCENTS
USA
0727

CREATIVE COMMUNE / URBAN ACCENTS
USA
0728

CREATIVE COMMUNE / URBAN ACCENTS
USA
0729

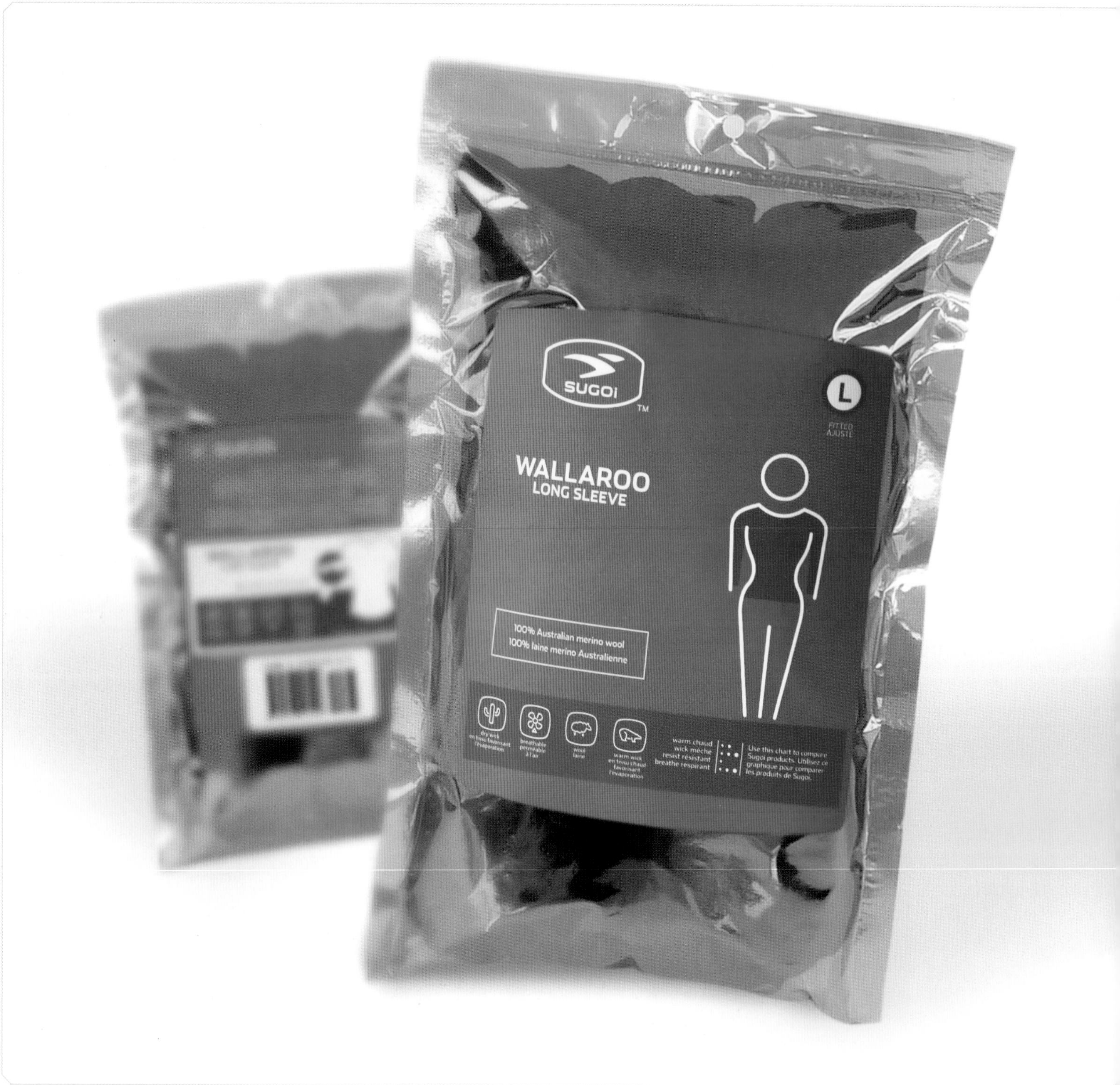
SUGOI
L
FITTED
AJUSTÉ
WALLAROO
LONG SLEEVE
100% Australian merino wool
100% laine merino Australienne
warm chaud
wick mèche
resist résistant
breathe respirant
Use this chart to compare Sugoi products. Utilisez ce graphique pour comparer les produits de Sugoi.

LEMLEY DESIGN COMPANY
USA

0730

LEMLEY DESIGN COMPANY
USA
0731

LEMLEY DESIGN COMPANY
USA
0732

LEMLEY DESIGN COMPANY
USA
0733

LEMLEY DESIGN COMPANY
USA
0734

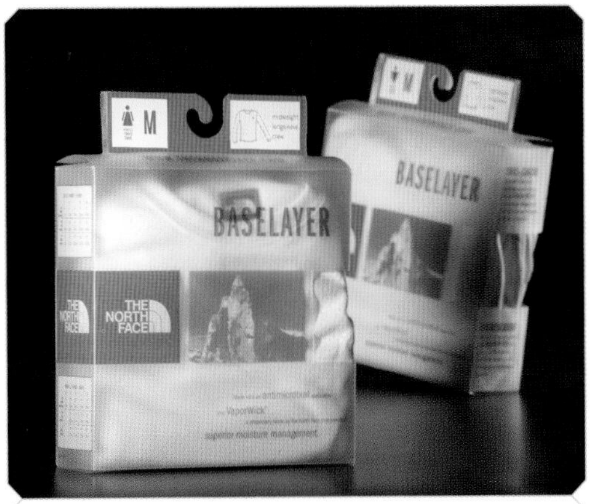

SATELLITE DESIGN
USA
0735

SATELLITE DESIGN
USA
0736

BLUE LOUNGE DESIGN
USA
0737

CONOVER
USA
0738

CREATIVE COMMUNE / URBAN ACCENTS
USA
0739

DEPERSICO CREATIVE GROUP
USA
0740

DEPERSICO CREATIVE GROUP
USA
0741

MICHAEL OSBORNE DESIGN
USA
0742

LLC "DESIGN DEPOT"
RUSSIA
0743

EVENSON DESIGN GROUP
USA
0744

LLC "DESIGN DEPOT"
RUSSIA
0745

KBDA
USA
0746

KBDA
USA
0747

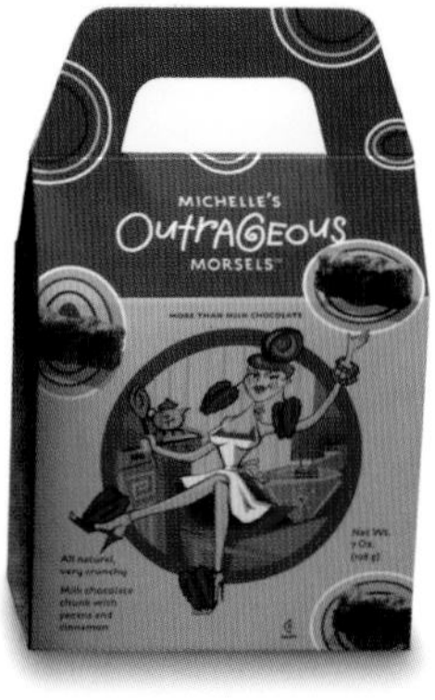

KBDA
USA
0748

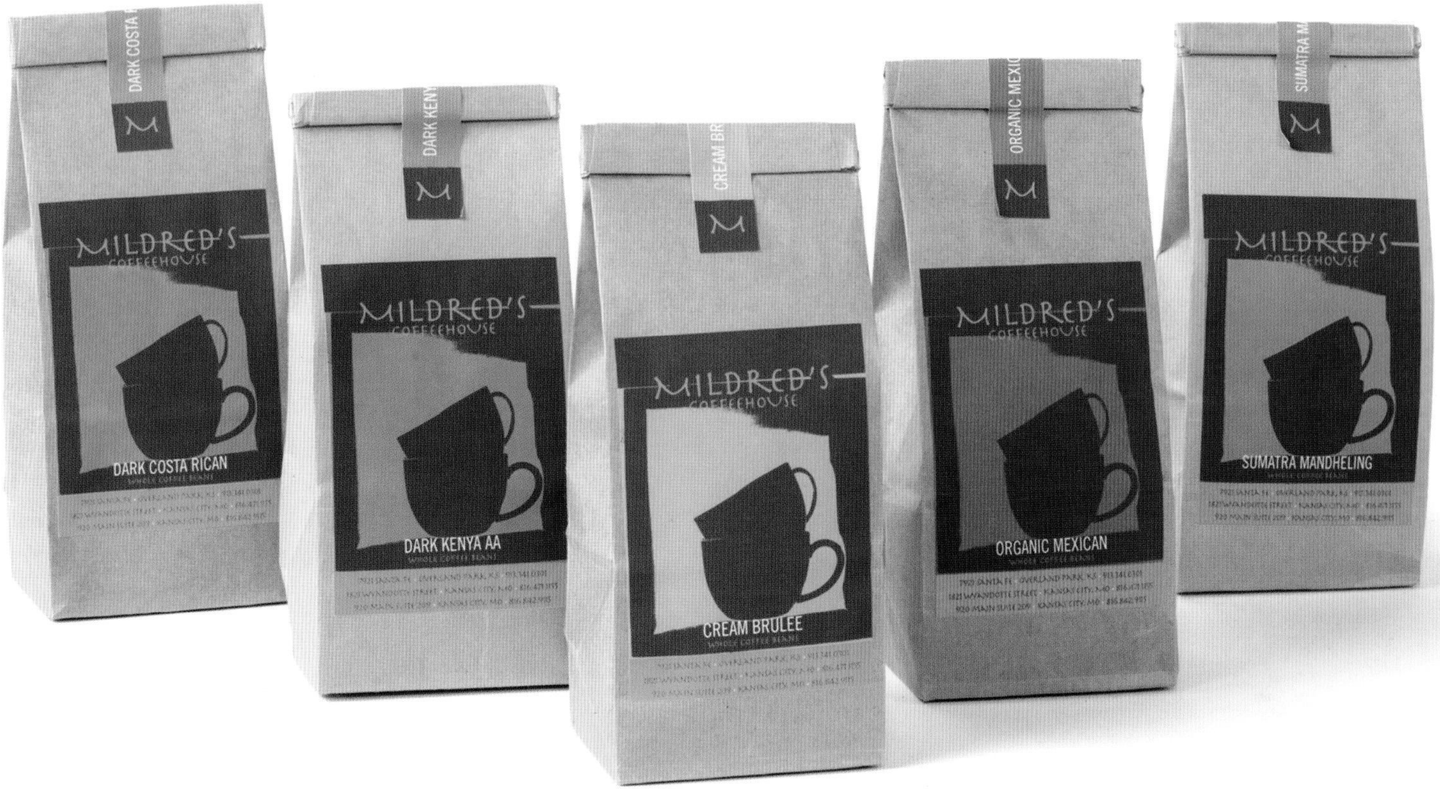
DARK COSTA
MILDRED'S
COFFEEHOUSE
DARK COSTA RICAN
DARK KENYA
DARK KENYA AA
CREAM BR
CREAM BRULEE
ORGANIC MEXIC
ORGANIC MEXICAN
SUMATRA M
SUMATRA MANDHELING

BUNGALOW CREATIVE
USA

0749

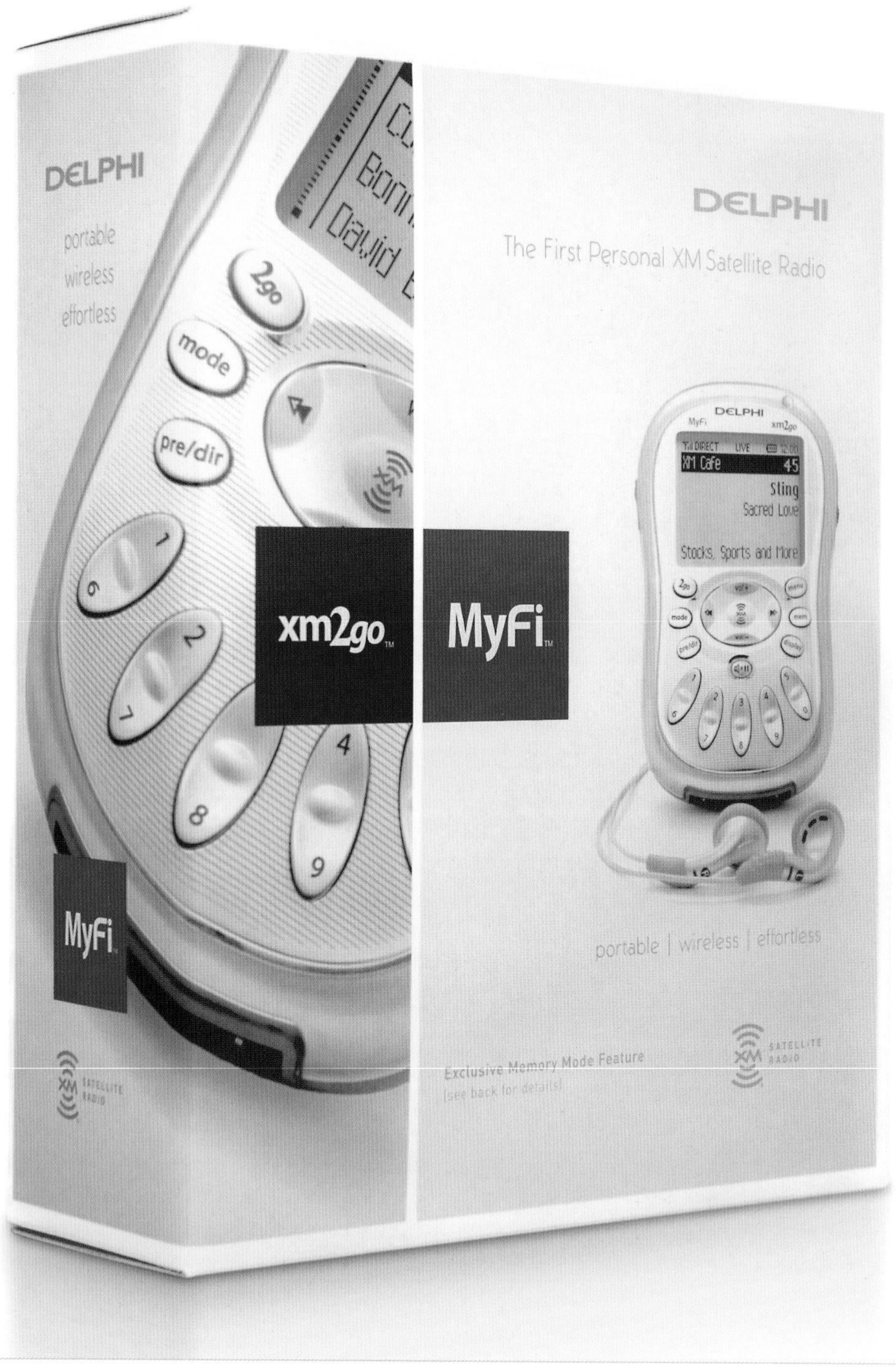

GO BIG
USA

0750

DELPHI
portable
wireless
effortless
2go
mode
pre/dir
xm2go
MyFi
MyFi
DELPHI
The First Personal XM Satellite Radio
XM Cafe
Sting
Sacred Love
Stocks, Sports and More
portable | wireless | effortless
Exclusive Memory Mode Feature
(see back for details)
XM SATELLITE RADIO

MILCH DESIGN GMBH
GERMANY
0751

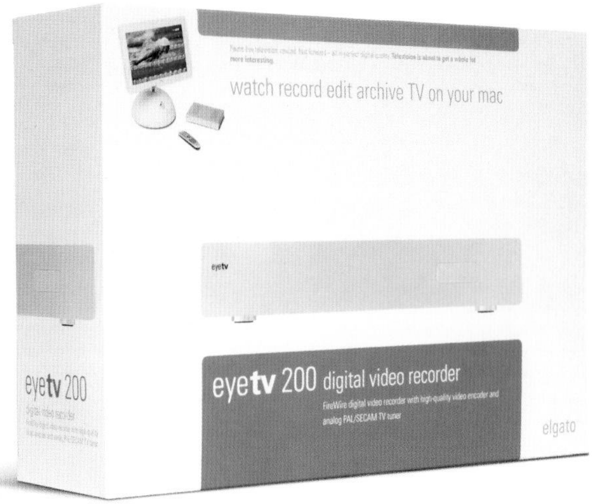

MILCH DESIGN GMBH
GERMANY
0752

MILCH DESIGN GMBH
GERMANY
0753

MILCH DESIGN GMBH
GERMANY
0754

CATALYST STUDIOS
USA
0755

PHILIPPE BECKER DESIGN
USA
0756

PHILIPPE BECKER DESIGN
USA
0757

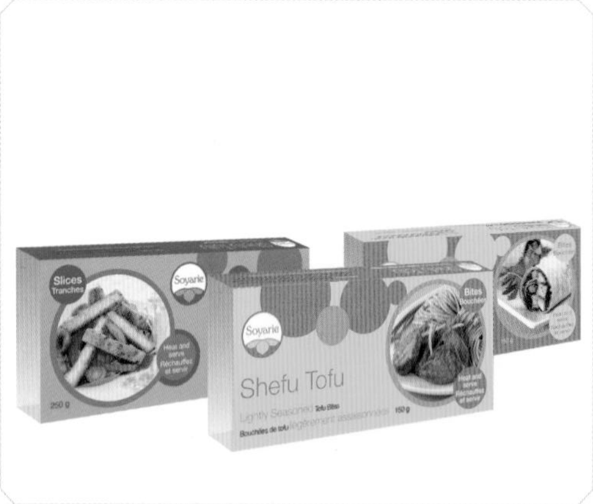

FIFTY STRATEGY & CREATIVE
CANADA
0758

FIFTY STRATEGY & CREATIVE
CANADA
0759

IRVING
UK
0760

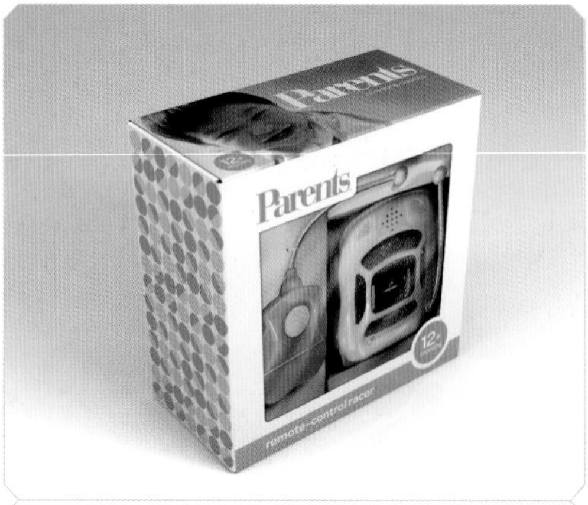

ON-PURPOS, INC.
USA
0761

MARY HUTCHISON DESIGN LLC
USA
0762

KENDALL ROSS
USA
0763

water • fruit • vitamins
iQ
H2O
green apple
lemon lime
mixed berry
16 FL OZ
473ML
orange mango
pomegranate

CRAVE, INC.
USA

0764

MOTOROLA
11/10/2006
11:35
Options
Main Menu
motorola
ROKR
E2

TURNER DUCKWORTH
USA

0765

TURNER DUCKWORTH
USA

0766

TURNER DUCKWORTH
USA

0767

TURNER DUCKWORTH
USA

0768

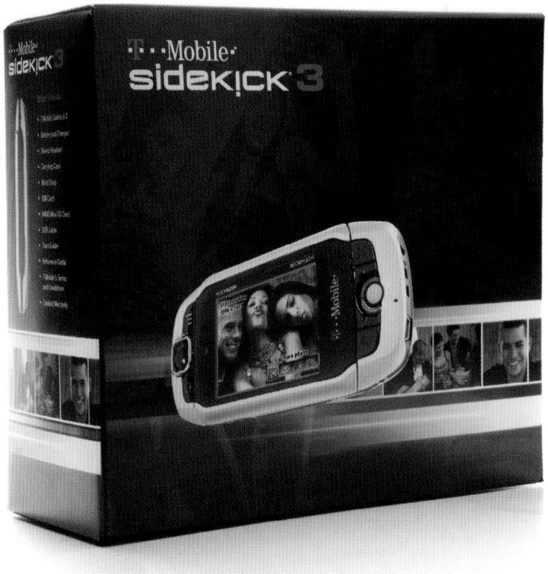

PHILIPPE BECKER DESIGN
USA

0769

MICHAEL OSBORNE DESIGN
USA
0770

THE MICRO AGENCY
CANADA
0771

THE MICRO AGENCY
CANADA
0772

SIBLEY PETEET DESIGN
USA
0773

SIBLEY PETEET DESIGN
USA
0774

SIBLEY PETEET DESIGN
USA
0775

CATO PARNELL
AUSTRALIA
0776

MICHAEL OSBORNE DESIGN
USA
0777

MILTON GLASER, INC.
USA
0778

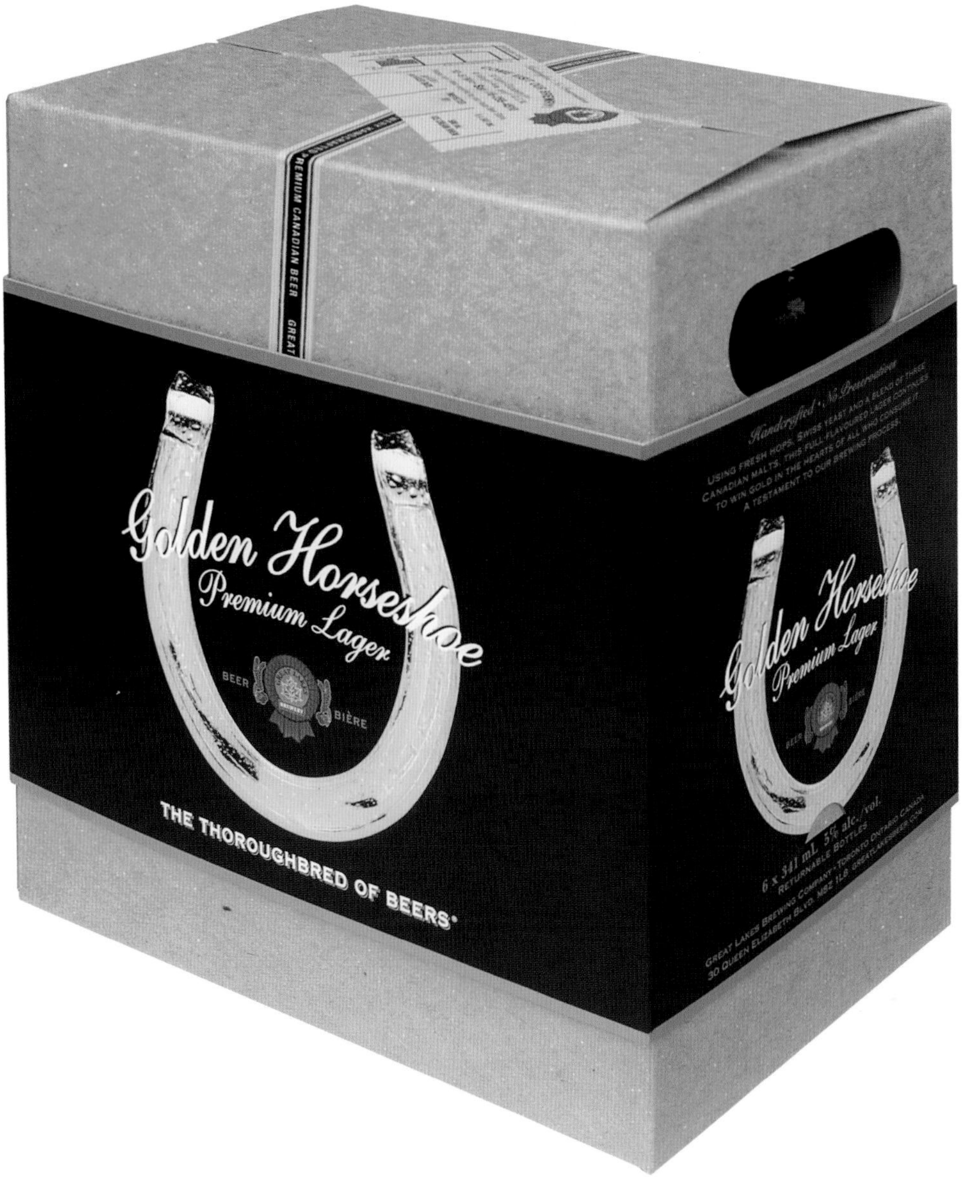

THE MICRO AGENCY
CANADA

0779

P&W
UK
0780

P&W
UK
0781

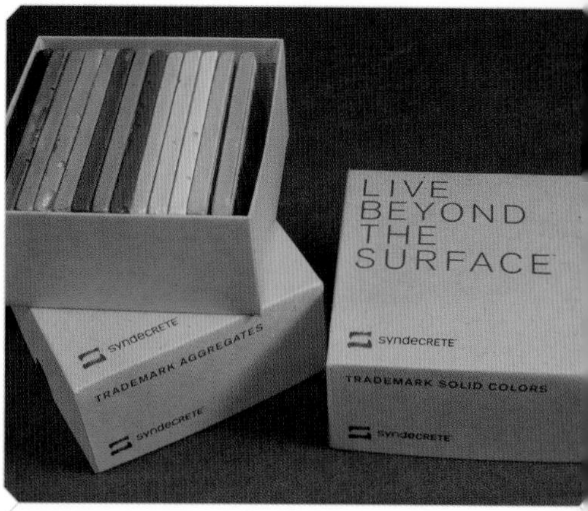

CONOVER
USA
0782

MICHAEL OSBORNE DESIGN
USA
0783

EVENSON DESIGN GROUP
USA
0784

MUKU STUDIOS, LLC
USA
0785

YELLOBEE STUDIO
USA
0786

YELLOBEE STUDIO
USA
0787

YELLOBEE STUDIO
USA
0788

LEMLEY DESIGN COMPANY
USA

0789

LLC "DESIGN DEPOT"
RUSSIA

0790

MICHAEL OSBORNE DESIGN
USA

0791

LEMLEY DESIGN COMPANY
USA

0792

HORNALL ANDERSON DESIGN
USA
0793

HORNALL ANDERSON DESIGN
USA
0794

HORNALL ANDERSON DESIGN
USA
0795

BAKKEN CREATIVE CO.
USA
0796

FULLBLASTINC.COM
USA
0797

FULLBLASTINC.COM
USA
0798

HORNALL ANDERSON DESIGN
USA
0799

HORNALL ANDERSON DESIGN
USA
0800

HORNALL ANDERSON DESIGN
USA
0801

CARTA d'IMBARCO
UNO DIREZIONE
ITALIA - CALIFORNIA
ZINFANDEL
DESTINAZIONE
tamás estates
VALITO PER UNA SOLA CORSA
ANNO
QUANTITÀ
ALC./VOL.
2004
750ML
13.5%
CARTA d'IMBARCO
UNO DIREZIONE
ITALIA - CALIFORNIA
PINOT GRIGIO
DESTINAZIONE
tamás estates
VALITO PER UNA SOLA CORSA
ANNO
QUANTITÀ
ALC./VOL.
2004
750ML
13.5%
CARTA d'IMBARCO
UNA DIREZIONE
ITALIA - CALIFORNIA
SANGIOVESE
DESTINAZIONE
tamás estates
VALITO PER UNA SOLA CORSA
ANNO
QUANTITÀ
ALC./VOL.
2004
750ML
13.5%

MICHAEL OSBORNE DESIGN
USA

0802

Zhena's Gypsy Tea
Full Leaf
Aromatherapeutic™
Iced Tea
Organic & Fair Trade
Mango Ginger White
White Tea
Exotic Mango, Ginger, Sweet Blackberry Leaf & Roses
5 1 Quart Silk Tea Pillows™ made from
100% biodegradable, GMO free, corn silk
Net Wt 1.75 oz (50g)

BLISS LANE STUDIOS
USA

0803

BLISS LANE STUDIOS
USA
0804

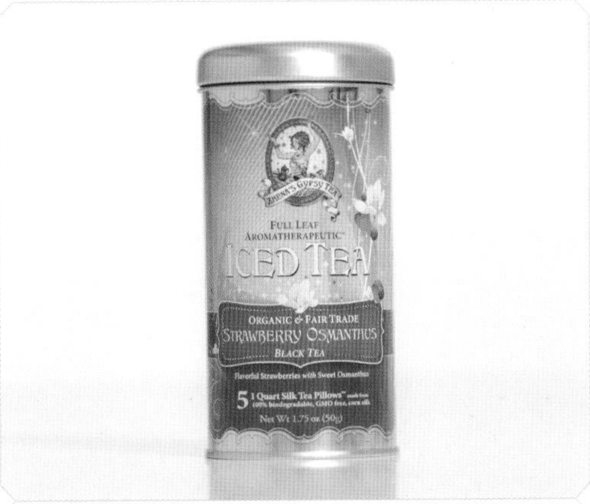

BLISS LANE STUDIOS
USA
0805

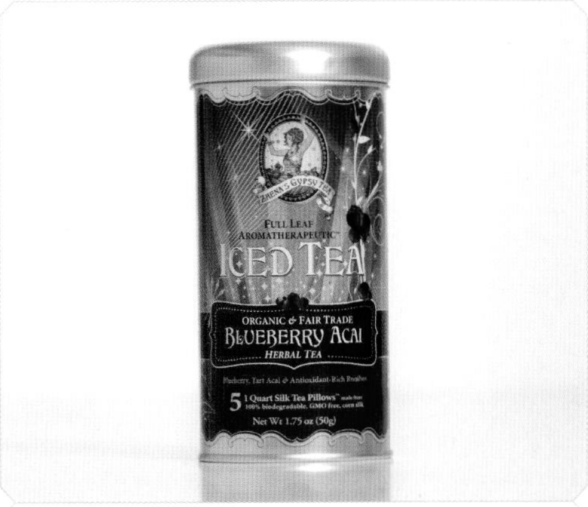

BLISS LANE STUDIOS
USA
0806

SIBLEY PETEET DESIGN
USA
0807

JOSHUA L. HUDSON
USA
0808

KBDA
USA
0809

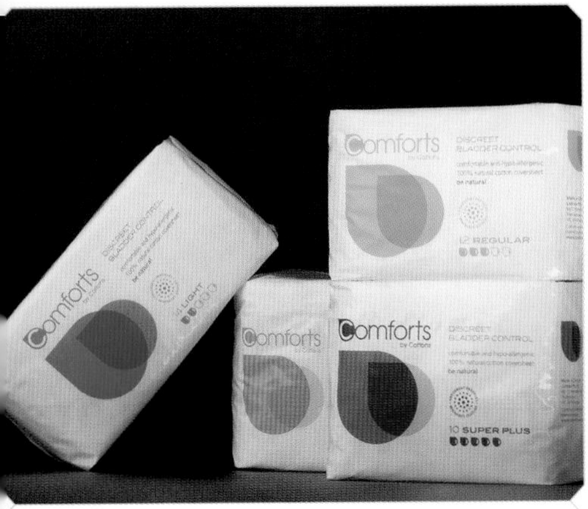

SPARK STUDIO
AUSTRALIA
0810

SPARK STUDIO
AUSTRALIA
0811

CATALYST STUDIOS
USA
0812

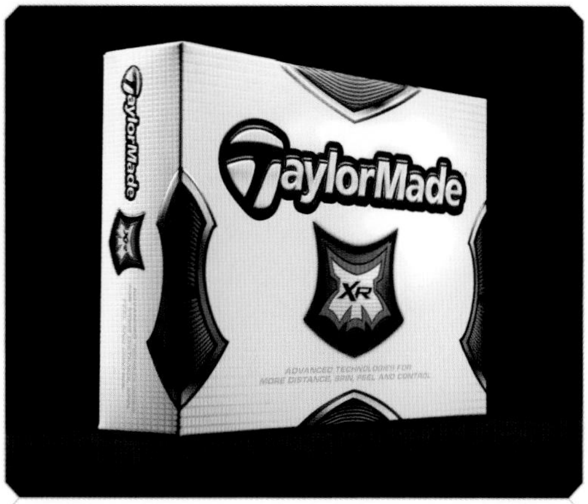

MIRIELLO GRAFICO
USA
0813

BONBON LONDON
UK
0814

DESIGN BRIDGE
UK
0815

DESIGN BRIDGE
UK
0816

RANDY MOSHER DESIGN
USA
0817

RANDY MOSHER DESIGN
USA
0818

THE O GROUP
USA
0819

WALLACE CHURCH, INC.
USA
0820

WALLACE CHURCH, INC.
USA
0821

S'ENIVRER
Cocktail Glasses
2 PC. Martini Set
The S'enivrer Martini Glass is a classic take on the most sophisticated cocktail. The impeccable clarity of the flawless crystal makes for an incredible tasting experience. The Martini stem is perfectly proportioned to hold 7.5 oz. straight up.

JOSHUA L. HUDSON
USA

0822

tire shine
Make rubber shine like new in seconds
Prevent fading and cracking
No more tired tires
vroom
16 FL OZ (1 PT) 473 mL
CAUTION: EYE IRRITANT. See Caution on back label.

TURNER DUCKWORTH
USA

0823

protectant wipes

Immediately restore that
new-car sheen

Prevent fading and cracking
from sun damage

Keep your car looking young

vroom

25 COUNT / 7"X 9" (17.7cm X 22.8cm)
CAUTION: EYE IRRITANT. See Caution on back label.

TURNER DUCKWORTH
USA

0824

wash me

microfiber sponge

Pamper your paintwork with a smooth
and sudsy wash

Great for polishing and dusting too

Touch me, i'm soft

vroom

8 3/4 in. x 4 3/4 in. x 2 in. (22 cm x 12 cm x 5 cm)

TURNER DUCKWORTH
USA

0825

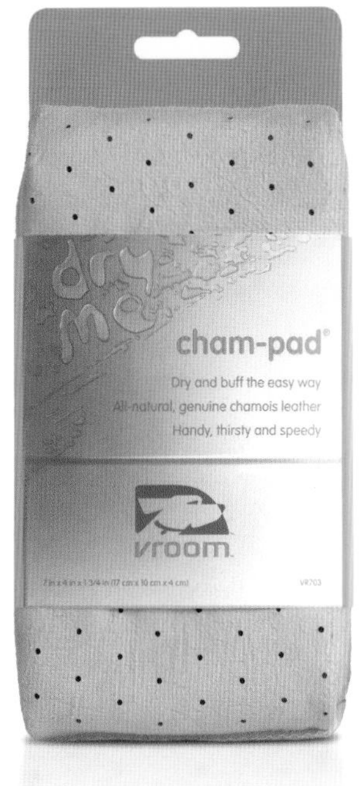

dry me

cham-pad®

Dry and buff the easy way

All-natural, genuine chamois leather

Handy, thirsty and speedy

vroom

7 in x 4 in x 1 3/4 in (17 cm x 10 cm x 4 cm)   VR703

TURNER DUCKWORTH
USA

0826

shine me

4 applicator pads

Effortlessly apply protectants,
polishes and waxes

Easy to hold and non-scratching

Handy, squishy and soft

vroom

4 pads

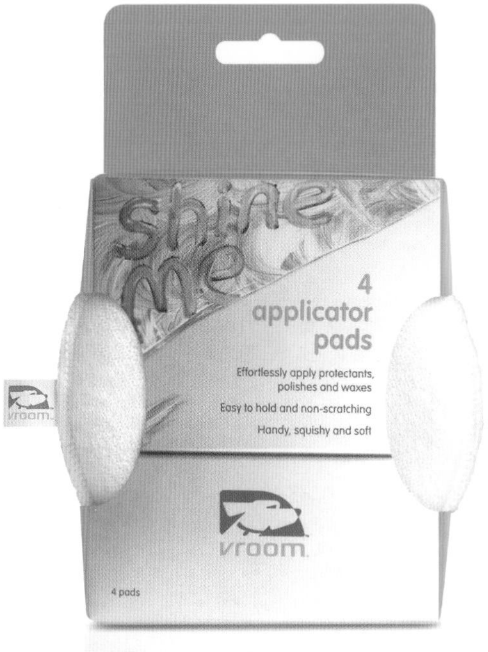

TURNER DUCKWORTH
USA

0827

P&W
UK
0828

P&W
UK
0829

P&W
UK
0830

MICHAEL OSBORNE DESIGN
USA
0831

KENDALL ROSS
USA
0832

TOKY BRANDING + DESIGN
USA
0833

P&W
UK
0834

P&W
UK
0835

P&W
UK
0836

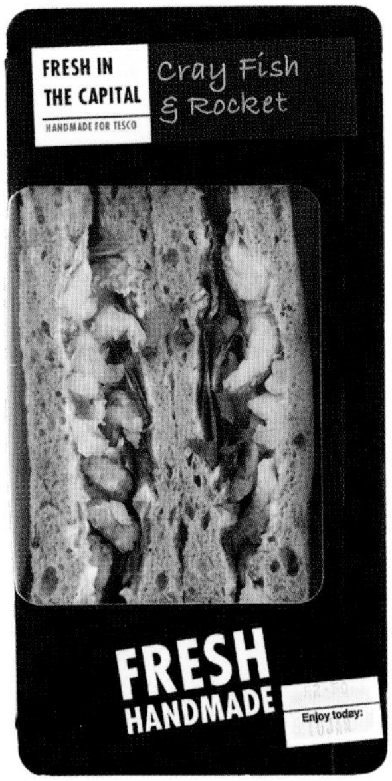

P&W
UK
0837

P&W
UK
0838

TURNER DUCKWORTH
USA
0839

P&W
UK
0840

neat-o!

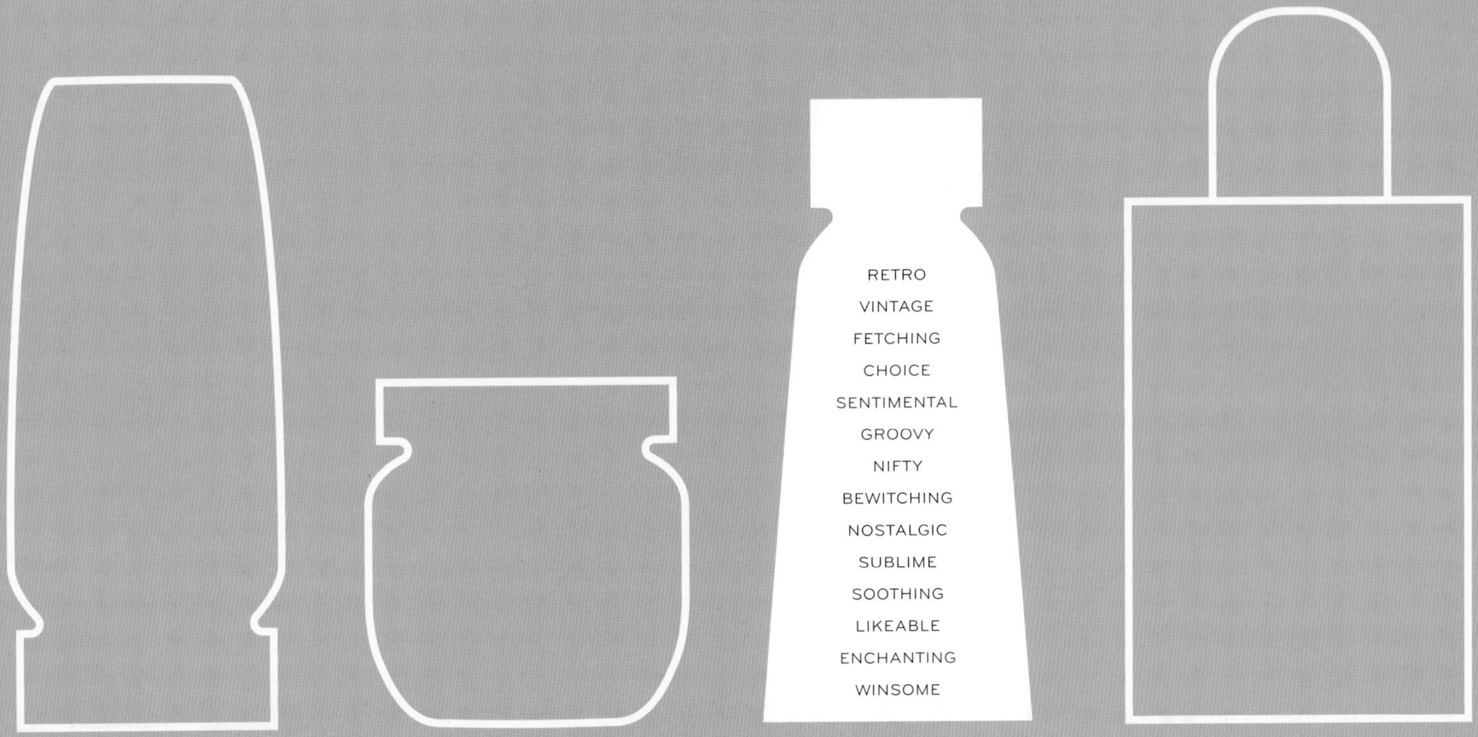
RETRO
VINTAGE
FETCHING
CHOICE
SENTIMENTAL
GROOVY
NIFTY
BEWITCHING
NOSTALGIC
SUBLIME
SOOTHING
LIKEABLE
ENCHANTING
WINSOME

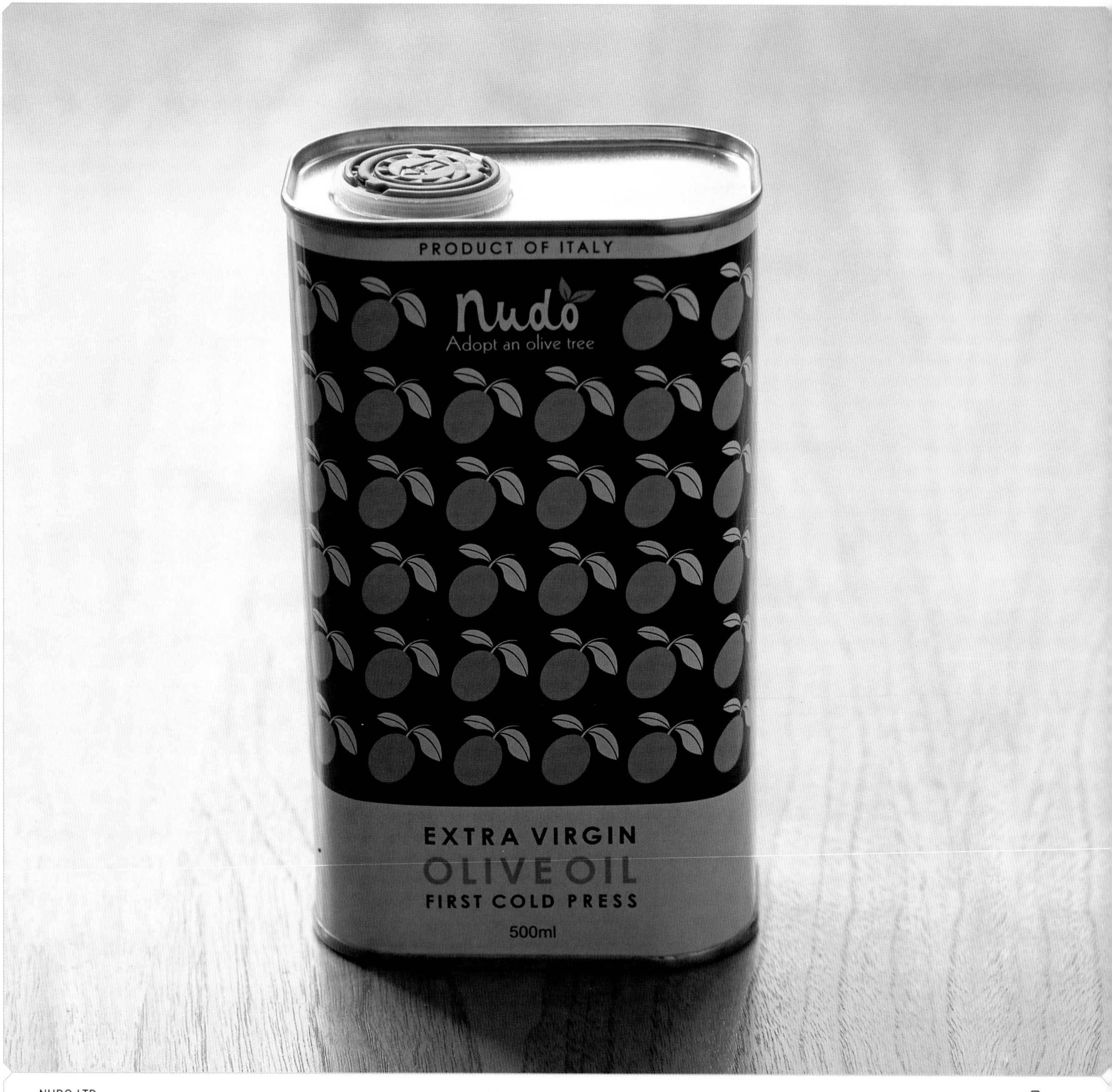
PRODUCT OF ITALY
Nudo
Adopt an olive tree
EXTRA VIRGIN
OLIVE OIL
FIRST COLD PRESS
500ml

NUDO LTD
UK

0841

OBJECT 9
USA

0842

LLOYD'S GRAPHIC DESIGN LTD.
NEW ZEALAND

0843

P&W
UK

0844

DESIGN BRIDGE
UK

0845

GO BIG
USA
0846

GO BIG
USA
0847

TURNER DUCKWORTH
USA
0848

TURNER DUCKWORTH
USA
0849

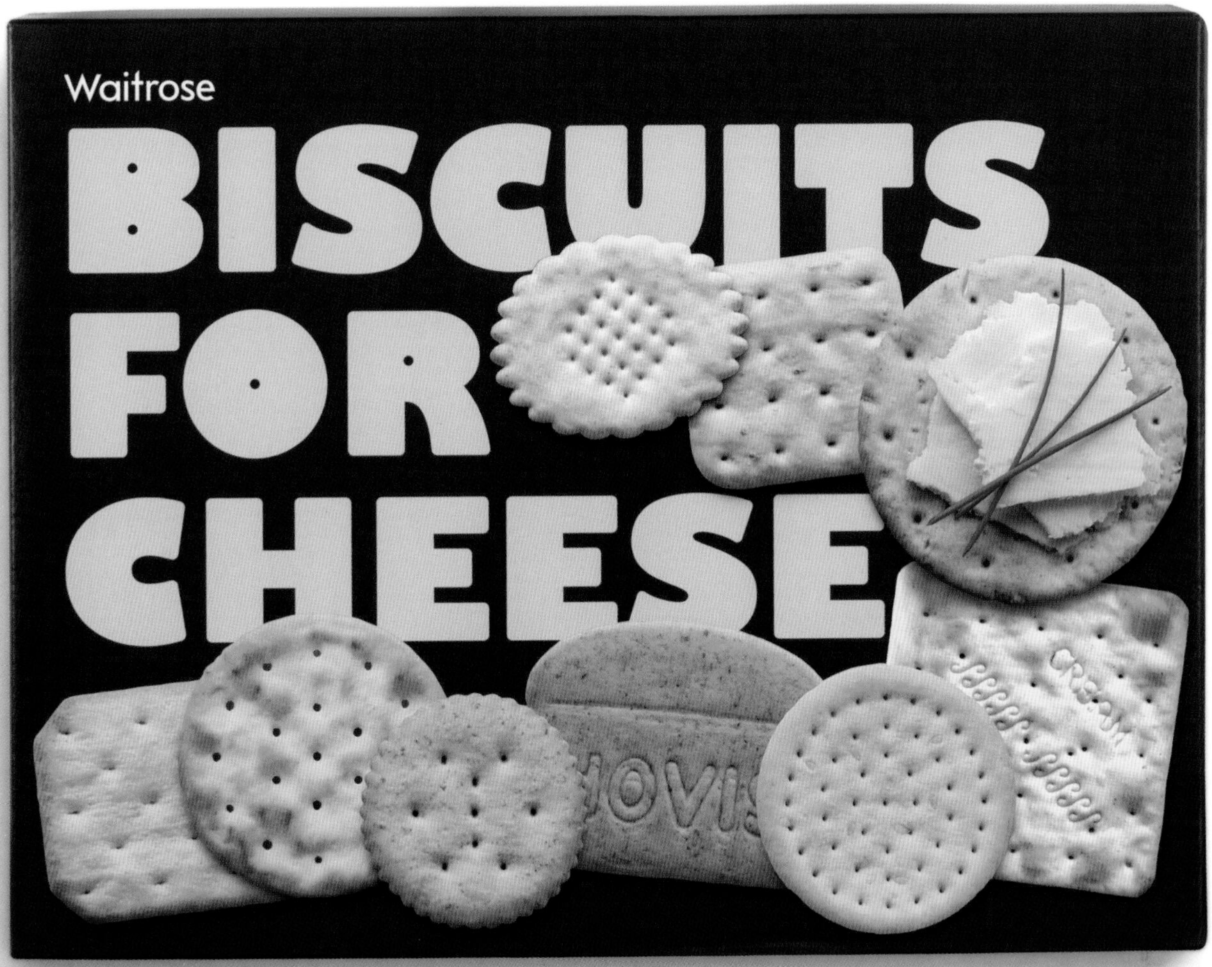
Waitrose
BISCUITS FOR CHEESE

TURNER DUCKWORTH
USA

0850

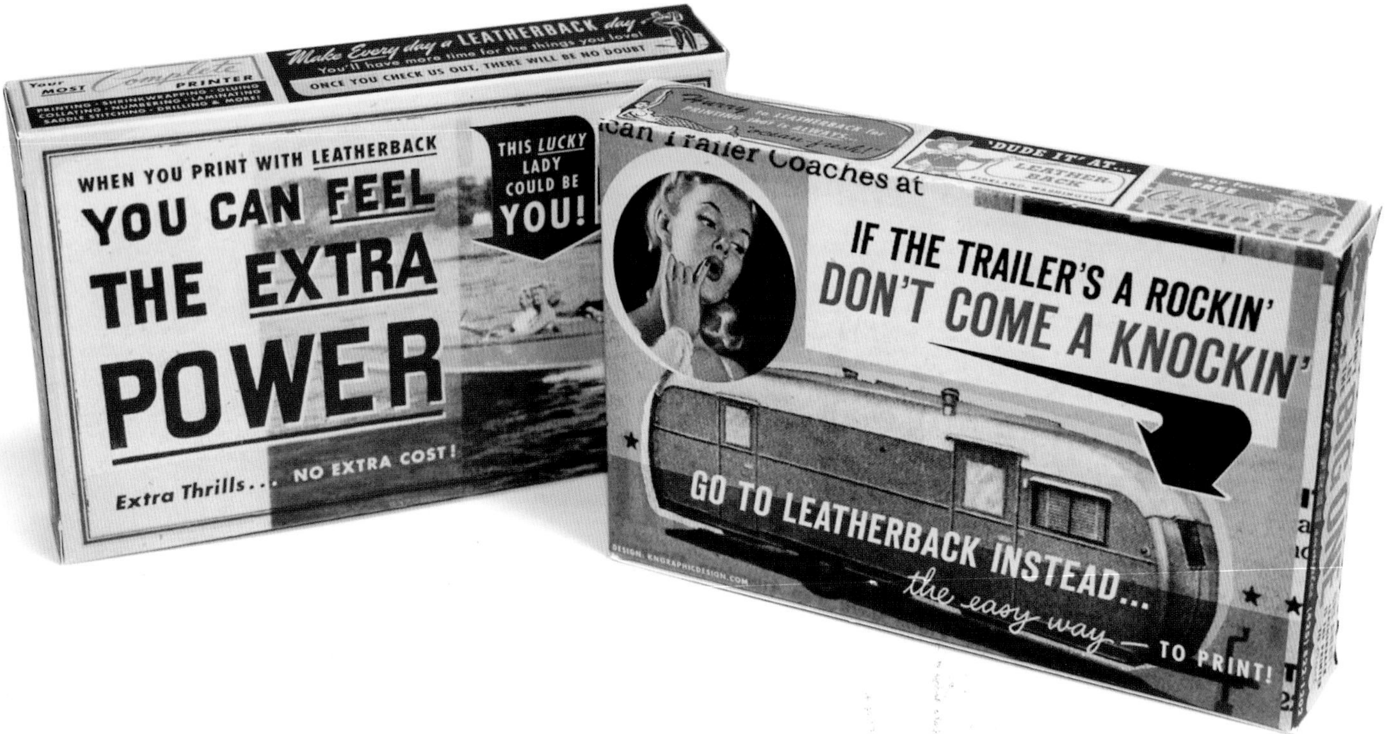
Make Every day a LEATHERBACK day
You'll have more time for the things you love!
ONCE YOU CHECK US OUT, THERE WILL BE NO DOUBT
PRINTER
WHEN YOU PRINT WITH LEATHERBACK
YOU CAN FEEL
THE EXTRA
POWER
THIS LUCKY LADY COULD BE YOU!
Extra Thrills . . . NO EXTRA COST!
Coaches at
IF THE TRAILER'S A ROCKIN'
DON'T COME A KNOCKIN'
GO TO LEATHERBACK INSTEAD...
the easy way — TO PRINT!

NIEDERMEIER
USA

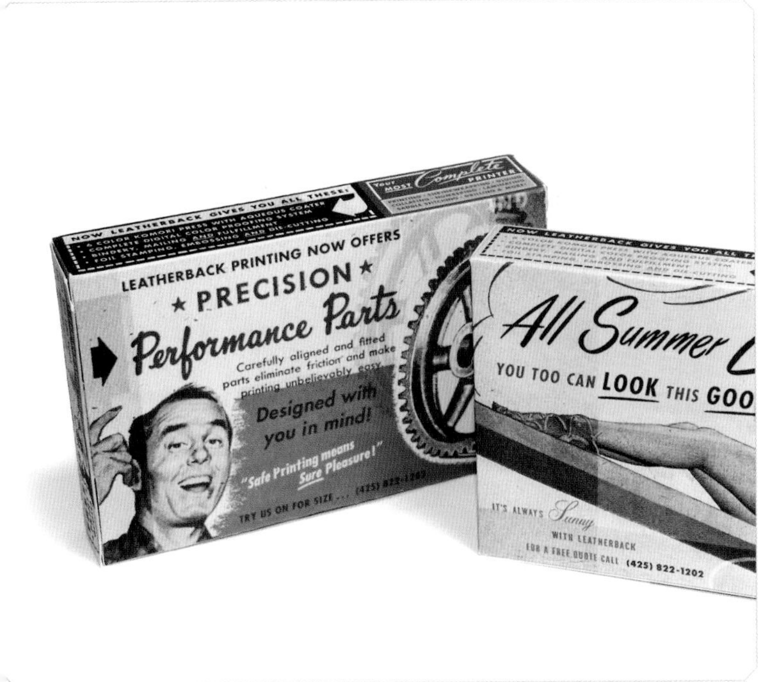
LEATHERBACK PRINTING NOW OFFERS
★ PRECISION ★
Performance Parts
Carefully aligned and fitted parts eliminate friction and make printing unbelievably easy.
Designed with you in mind!
"Safe Printing means Sure Pleasure!"
TRY US ON FOR SIZE ... (425) 822-1202
All Summer
YOU TOO CAN LOOK THIS GOO
IT'S ALWAYS Sunny
WITH LEATHERBACK
FOR A FREE QUOTE CALL (425) 822-1202

NIEDERMEIER
USA

0852

Black Eye Design
KNOCKOUT!
WHISKEY
GINGER
314 mL
fine shred marmalade

BLACK EYE DESIGN
CANADA

0853

POPE CREEK
RANCH
EXTRA VIRGIN
OLIVE OIL
EARLY HARVEST

MYINT DESIGN
USA

0854

Your ALL WEATHER PRINTER
LEATHERBACK day
ave more time for the things you love!
CHECK US OUT, THERE WILL BE NO DOUBT
Summers can be extremely ho
...but Leatherback is VERY CO
BEAT the HEAT!
'DUDE IT' AT...
LEATHER-BACK
KIRKLAND, WASHINGTON
Stop by for...
FREE
Colorful
SAMPLES!
DON'T LET THE BIG ON

NIEDERMEIER
USA

0855

MUCCA DESIGN CORP.
USA
0856

MUCCA DESIGN CORP.
USA
0857

MUCCA DESIGN CORP.
USA
0858

MUCCA DESIGN CORP.
USA
0859

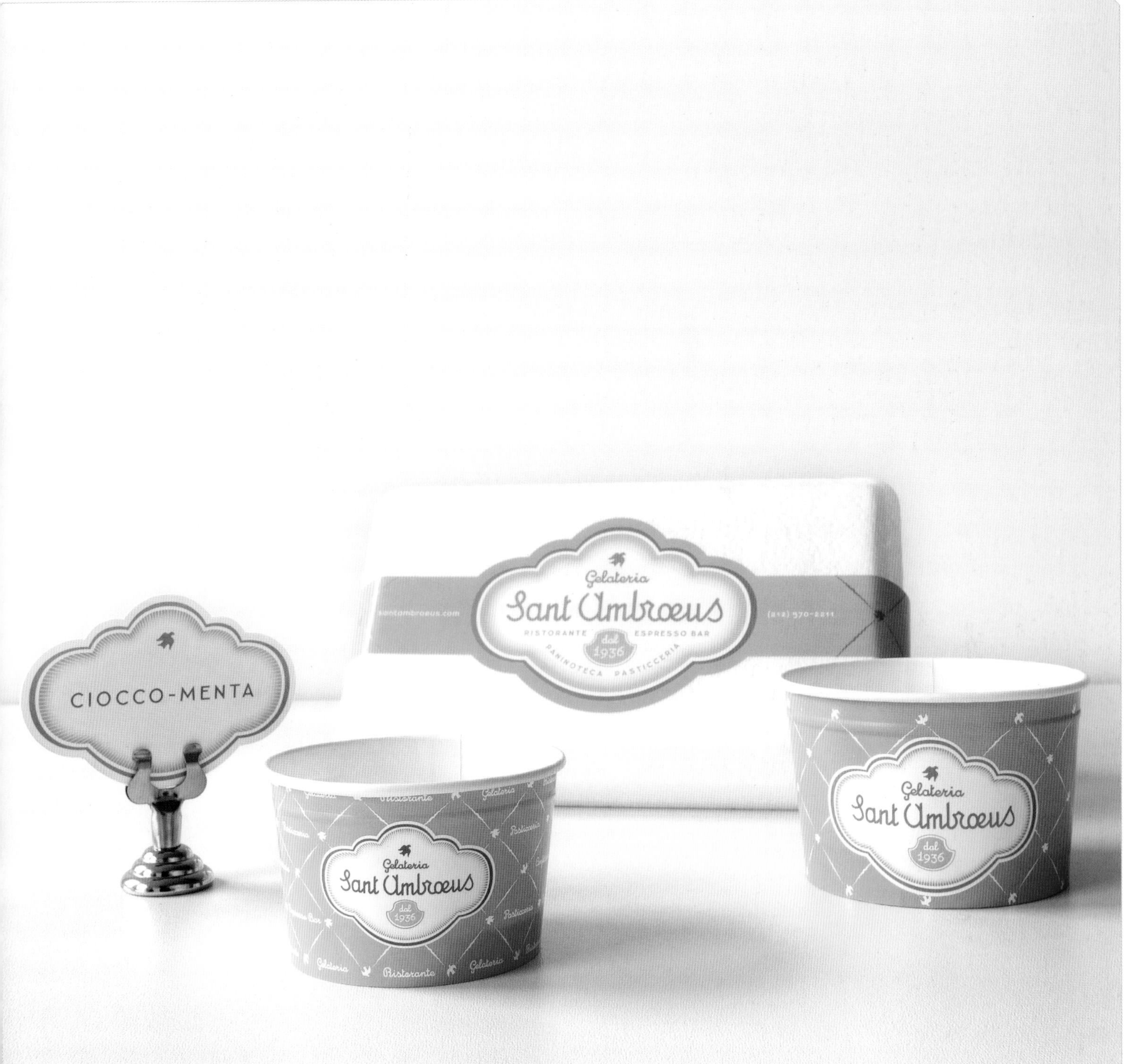
CIOCCO-MENTA
Gelateria
Sant Ambroeus
RISTORANTE
ESPRESSO BAR
dal 1936
PANINOTECA
PASTICCERIA

MUCCA DESIGN CORP.
USA

0860

TULLY'S
COFFEE
HOLIDAY
COAST
2004
GROUND
TULLY'S
COFFEE
DISCOVER
HOLIDAY
CHOCOLATE COVERED
CRANBERRIES
TRADITIONAL
MULLING
SPICES
DISCOVER
HOLIDAY
WITH US

LEMLEY DESIGN COMPANY
USA

0861

LEMLEY DESIGN COMPANY
USA
0862

LEMLEY DESIGN COMPANY
USA
0863

PHILIPPE BECKER DESIGN
USA
0864

MICHAEL OSBORNE DESIGN
USA
0865

TURNER DUCKWORTH
USA

0866

TURNER DUCKWORTH
USA

0867

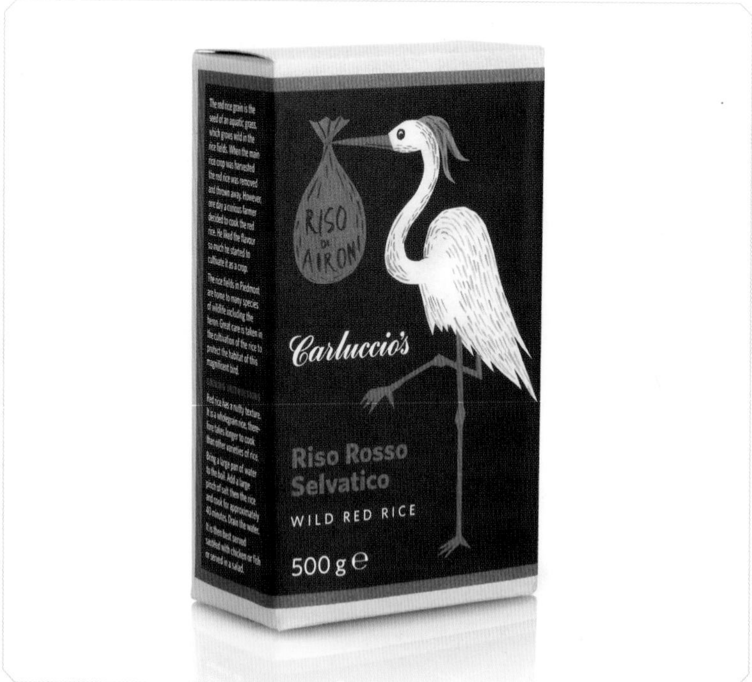

IRVING
UK

0868

MUCCA DESIGN CORP.
USA

0869

BALTHAZAR
FINE PORCELAIN
HOLDS 2 OZ.
A PAIR OF
ESPRESSO
CUPS AND SAUCERS
CRAFTED IN ITALY
HIGHEST QUALITY
80 SPRING STREET—NEW YORK

MUCCA DESIGN CORP.
USA

0870

Eckert's
COUNTRY STORE
SINCE 1837
Eckert's
Always in Season
APPLE
NET WT 2 LB 4 OZ
Eckert's
Always in Season
APPLE
Butter
REFRIGERATE AFTER OPENING
NET WT 1 LB 2 OZ (510g) • ONE PINT
Eckert's
NO SUGAR ADDED
APPLE
Butter
REFRIGERATE AFTER OPENING
NET WT 1 LB 2 OZ (510g) • ONE PINT
Eckert's
APPLE
Butter

TOKY BRANDING + DESIGN
USA

0871

PHILIPPE BECKER DESIGN
USA
0872

TOKY BRANDING + DESIGN
USA
0873

THYNK DESIGN, INC.
USA
0874

PHILIPPE BECKER DESIGN
USA
0875

TOKY BRANDING + DESIGN
USA

0876

PHILIPPE BECKER DESIGN
USA

0877

PHILIPPE BECKER DESIGN
USA

0878

TOKY BRANDING + DESIGN
USA

0879

Eckert's
Always in Season
Pure
HONEY
NET WT
PACKED FOR
ECKERT'S COUNTRY STORE
BELLEVILLE, ILL.
Eckert's
Always in Season
Pure
HONEY
NET WT
PACKED FOR
ECKERT'S COUNTRY STORE
BELLEVILLE, ILL.
Eckert's
Always in Season
Pure
HONEY
NET WT
PACKED FOR
ECKERT'S COUNTRY STORE
BELLEVILLE, ILL.
Eckert's
Always in Season
Comb
HONEY
NET WT

TOKY BRANDING + DESIGN
USA

o88o

LIMONCELLO
LIMONCELLO
ALCOHOL 32% VOL.

BAKKEN CREATIVE
USA

0881

MICHAEL SCHWAB STUDIO
USA
0882

MICHAEL SCHWAB STUDIO
USA
0883

MICHAEL SCHWAB STUDIO
USA
0884

MICHAEL SCHWAB STUDIO
USA
0885

MICHAEL SCHWAB STUDIO
USA
0886

MICHAEL SCHWAB STUDIO
USA
0887

MICHAEL SCHWAB STUDIO
USA
0888

MICHAEL SCHWAB STUDIO
USA
0889

MICHAEL SCHWAB STUDIO
USA
0890

TURNER DUCKWORTH
USA
0891

TURNER DUCKWORTH
USA
0892

ELEMENT
USA
0893

ELEMENT
USA
0894

ELEMENT
USA
0895

ELEMENT
USA
0896

TURNER DUCKWORTH
USA
0897

TURNER DUCKWORTH
USA
0898

TURNER DUCKWORTH
USA
0899

Waitrose
RUBY
ORANGE &
PASSION FRUIT
CRUSH

TURNER DUCKWORTH
USA

0900

Holiday
TREATS KIT
Recipes and Wrappings for
the Tastiest Gifts of the Season

CHRONICLE BOOKS
USA

0901

CHRONICLE BOOKS
USA
0902

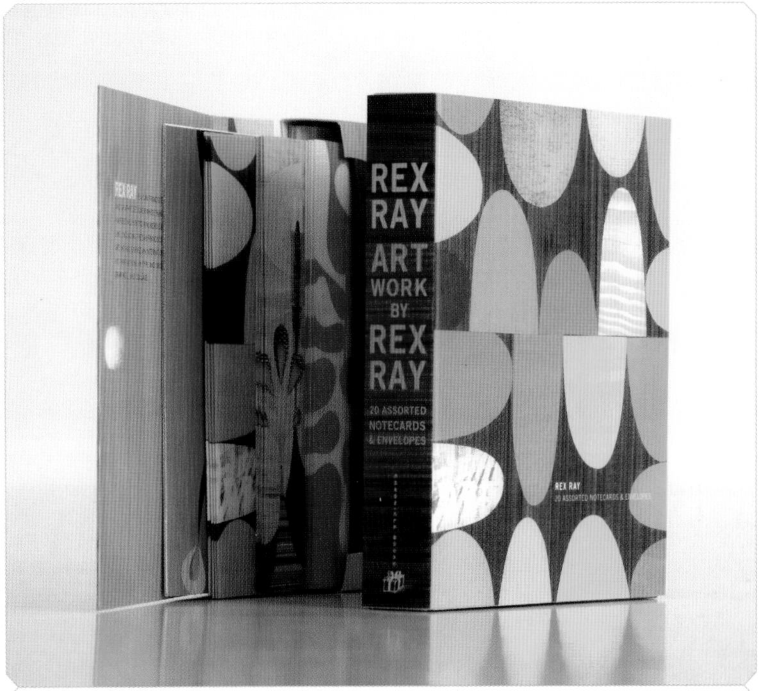

CHRONICLE BOOKS
USA
0903

CHRONICLE BOOKS
USA
0904

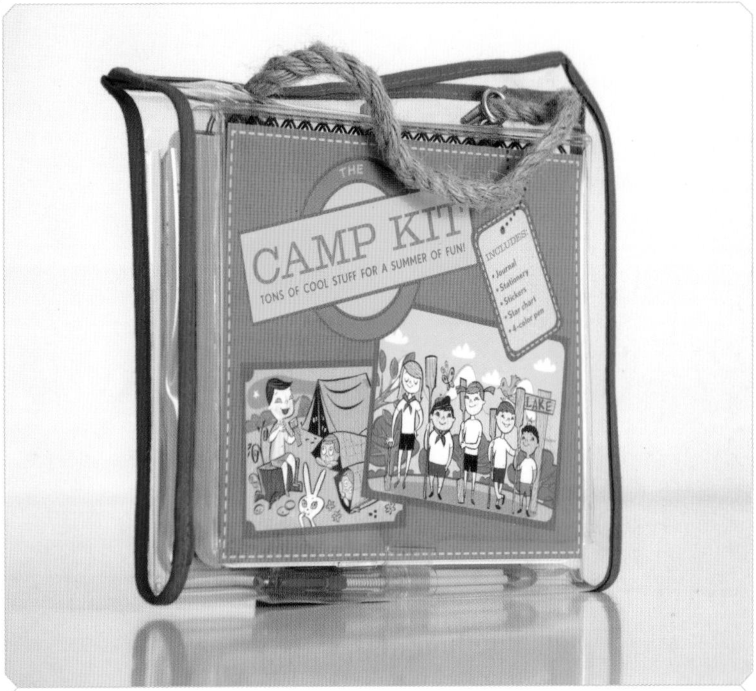

CHRONICLE BOOKS
USA
0905

BURGEFF CO.
MEXICO
0906

CAZEE DESIGN
USA
0907

TOKY BRANDING + DESIGN
USA
0908

TOKY BRANDING + DESIGN
USA
0909

CREATIVE COMMUNE / URBAN ACCENTS
USA
0910

CREATIVE COMMUNE / URBAN ACCENTS
USA
0911

MICHAEL OSBORNE DESIGN
USA
0912

TOKY BRANDING + DESIGN
USA
0913

TOKY BRANDING + DESIGN
USA
0914

BREWED IN
COLUMBUS, OHIO
Est 1988
Columbus
BREWING COMPANY
PALE ALE
COPPER-COLORED WITH A DRY FINISH

BREWED IN
COLUMBUS, OHIO
Est 1988
Columbus
BREWING COMPANY
90 SHILLING ALE
DARK, RUBY RED, SCOTTISH STYLE ALE

ELEMENT
USA

0915

BREWED IN
COLUMBUS, OHIO
Est 1988
Columbus
BREWING COMPANY
OHIO HONEY WHEAT
REFRESHINGLY SMOOTH, CRISP, BLONDE ALE

BREWED IN
COLUMBUS, OHIO
Est 1988
Columbus
BREWING COMPANY
APRICOT ALE
GOLDEN-COPPER, MEDIUM BODIED ALE

ELEMENT
USA

0916

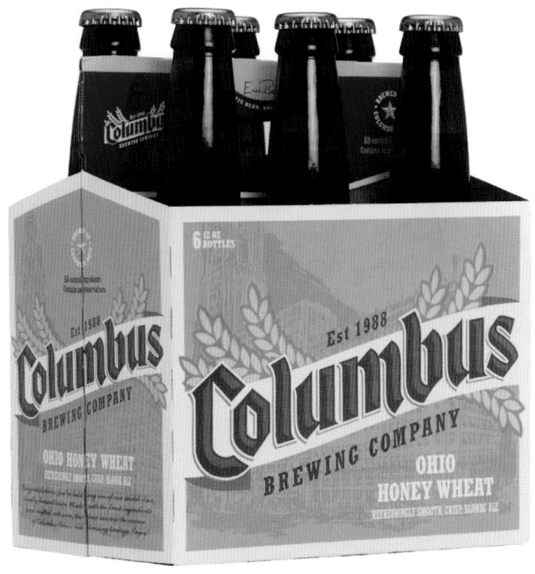

ELEMENT
USA

0917

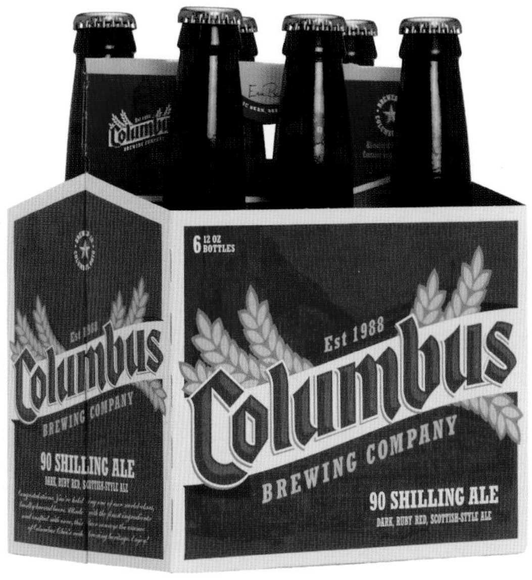

ELEMENT
USA

0918

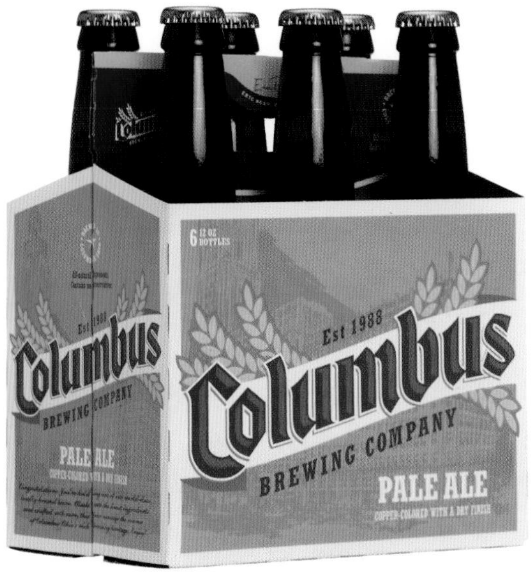

ELEMENT
USA

0919

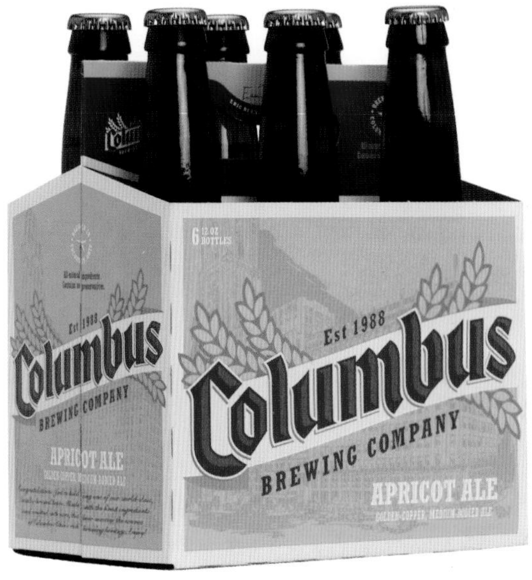

ELEMENT
USA

0920

WALLACE CHURCH, INC.
USA
0921

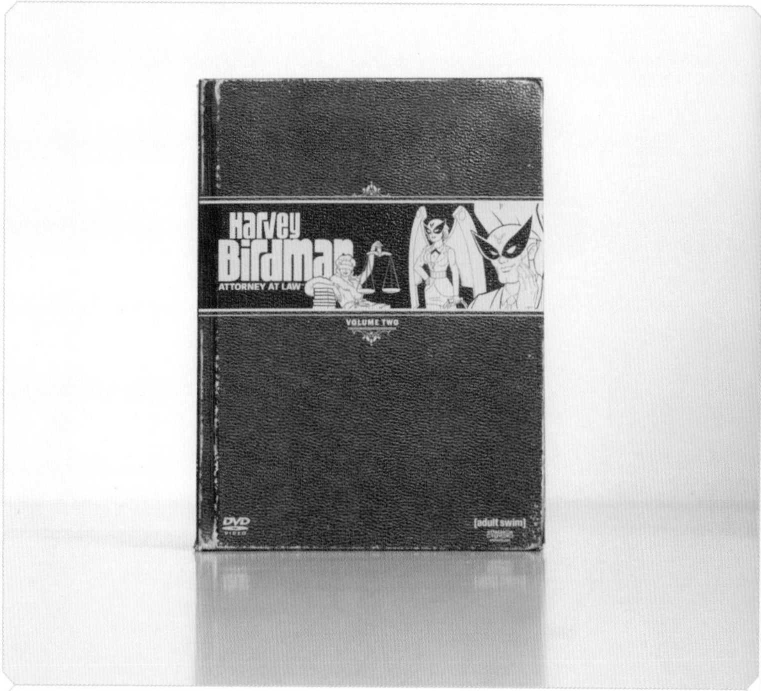

BAD STUDIO
USA
0922

BAD STUDIO
USA
0923

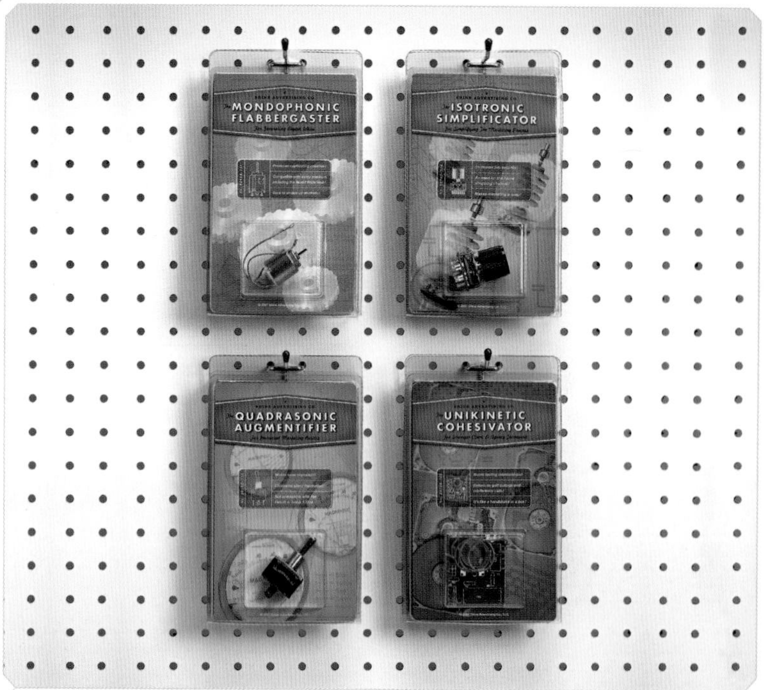

SHINE ADVERTISING
USA
0924

Harvey
Birdman
ATTORNEY AT LAW
VOLUME ONE
DVD
VIDEO
[adult swim]

BAD STUDIO
USA

0925

Rhino
BREW
Charge your glasses!
330ml
Charge your glasses!
Rhino
BREW
WhiteRhino

WHITERHINO CREATIVE
AUSTRALIA

0926

KENDALL ROSS
USA
0927

KENDALL ROSS
USA
0928

KENDALL ROSS
USA
0929

SOCKEYE CREATIVE
USA
0930

SOCKEYE CREATIVE
USA
0931

SOCKEYE CREATIVE
USA
0932

RANDY MOSHER DESIGN
USA
0933

DAVID CARTER DESIGN ASSOCIATES
USA
0934

TURNER DUCKWORTH
USA
0935

HOMART
USA
0936

HOMART
USA
0937

HOMART
USA
0938

BURGEFF CO.
MEXICO
0939

FUNNEL: ERIC KASS: UTILITARIAN + COMMERCIAL + FINE: ART
USA

0940

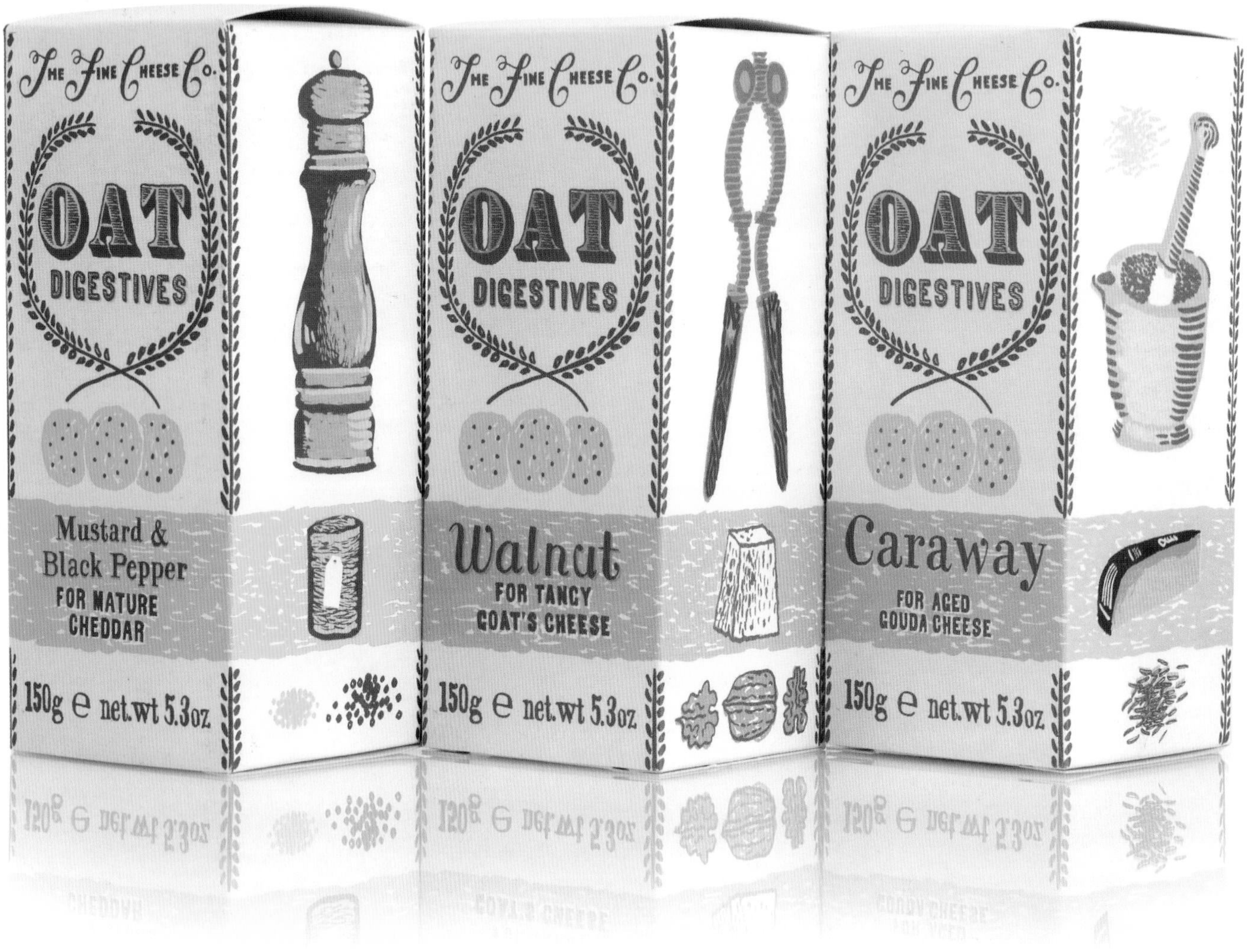
The Fine Cheese Co.
OAT
DIGESTIVES
Mustard & Black Pepper
FOR MATURE CHEDDAR
150g e net.wt 5.3oz
The Fine Cheese Co.
OAT
DIGESTIVES
Walnut
FOR TANGY GOAT'S CHEESE
150g e net.wt 5.3oz
The Fine Cheese Co.
OAT
DIGESTIVES
Caraway
FOR AGED GOUDA CHEESE
150g e net.wt 5.3oz

IRVING
UK

0941

O! ADVERTISING & DESIGN
ICELAND
0942

PHILIPPE BECKER DESIGN
USA
0943

SAGMEISTER INC.
USA
0944

O! ADVERTISING & DESIGN
ICELAND
0945

THE DECODER RING DESIGN CONCERN
USA
0946

DECARLO DESIGN CO.
USA
0947

EPOS, INC.
USA
0948

BAD STUDIO
USA
0949

BAD STUDIO
USA
0950

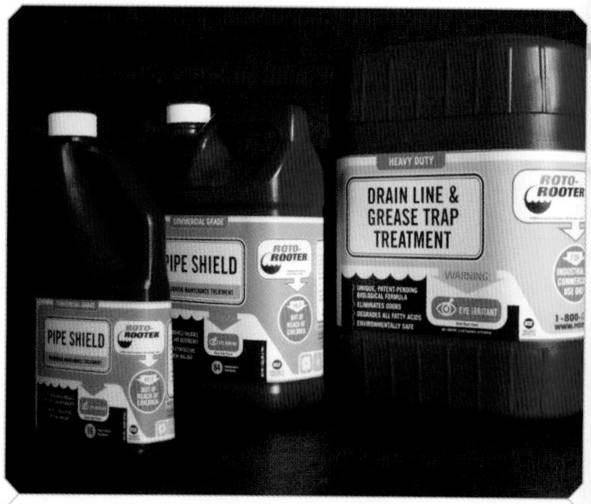

SAYLES GRAPHIC DESIGN
USA
0951

PEACE COFFEE
USA
0952

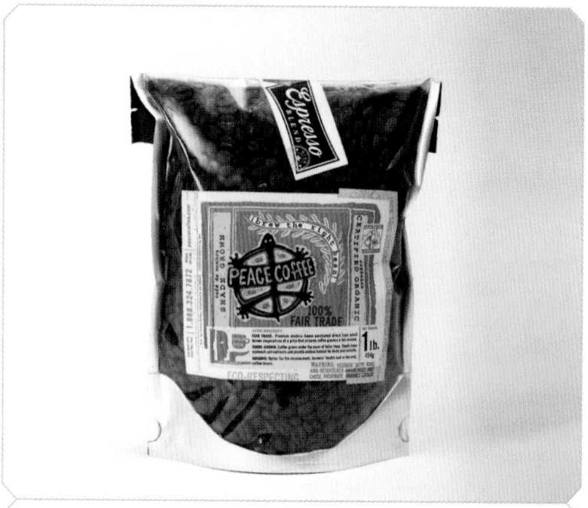

PEACE COFFEE
USA
0953

PEACE COFFEE
USA
0954

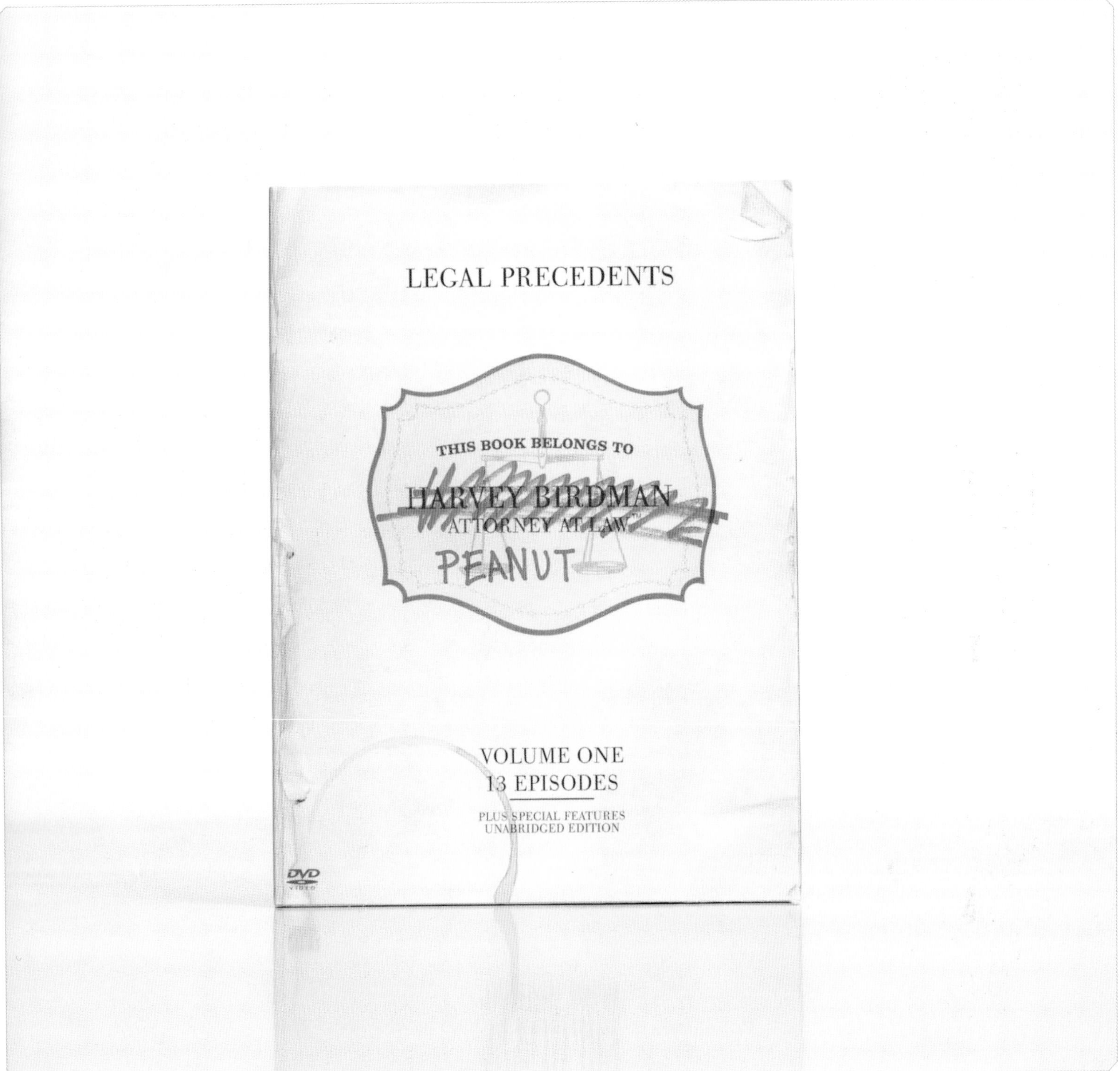
LEGAL PRECEDENTS
THIS BOOK BELONGS TO
HARVEY BIRDMAN
ATTORNEY AT LAW™
PEANUT
VOLUME ONE
13 EPISODES
PLUS SPECIAL FEATURES
UNABRIDGED EDITION
DVD

BAD STUDIO
USA

0955

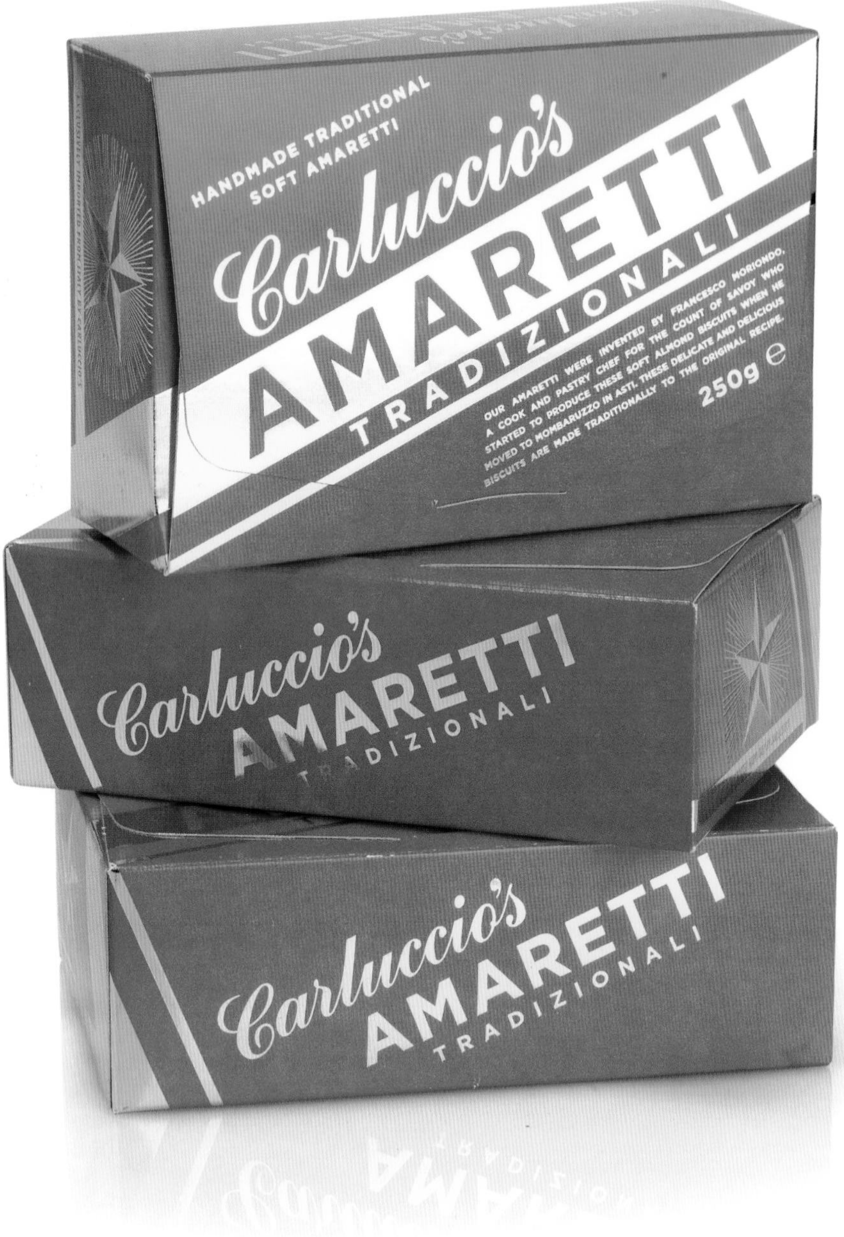
HANDMADE TRADITIONAL
SOFT AMARETTI
Carluccio's
AMARETTI
TRADIZIONALI
OUR AMARETTI WERE INVENTED BY FRANCESCO MORIONDO, A COOK AND PASTRY CHEF FOR THE COUNT OF SAVOY WHO STARTED TO PRODUCE THESE SOFT ALMOND BISCUITS WHEN HE MOVED TO MOMBARUZZO IN ASTI. THESE DELICATE AND DELICIOUS BISCUITS ARE MADE TRADITIONALLY TO THE ORIGINAL RECIPE.
250g ℮
Carluccio's
AMARETTI
TRADIZIONALI
Carluccio's
AMARETTI
TRADIZIONALI

IRVING
UK

0956

IRVING
UK
0957

IRVING
UK
0958

IRVING
UK
0959

IRVING
UK
0960

SUBPLOT DESIGN INC.
CANADA

0961

SUBPLOT DESIGN INC.
CANADA

0962

SUBPLOT DESIGN INC.
CANADA

0963

SUBPLOT DESIGN INC.
CANADA

0964

SAINT CLAIR
MARLBOROUGH
VARIETY: Sauvignon Blanc
YEAR: 2005
BLOCK: One
WINEMAKER:
NEW ZEALAND

THE CREATIVE METHOD
AUSTRALIA

0965

PRODUCT OF ITALY
Sant Ambroeus
Extra Virgin
Olive Oil
OLIO
di
OLIVA
Extravergine
produced by
FATTORIA
SARDI GIUSTINIANI
and bottled by
FRANTOIO
DEL COMPITESE
in Lucca, Italy
NET CONTENTS
25.4 FL OZ (1 PT 9.4 FL OZ) 750 ML

MUCCA DESIGN CORP.
USA

0966

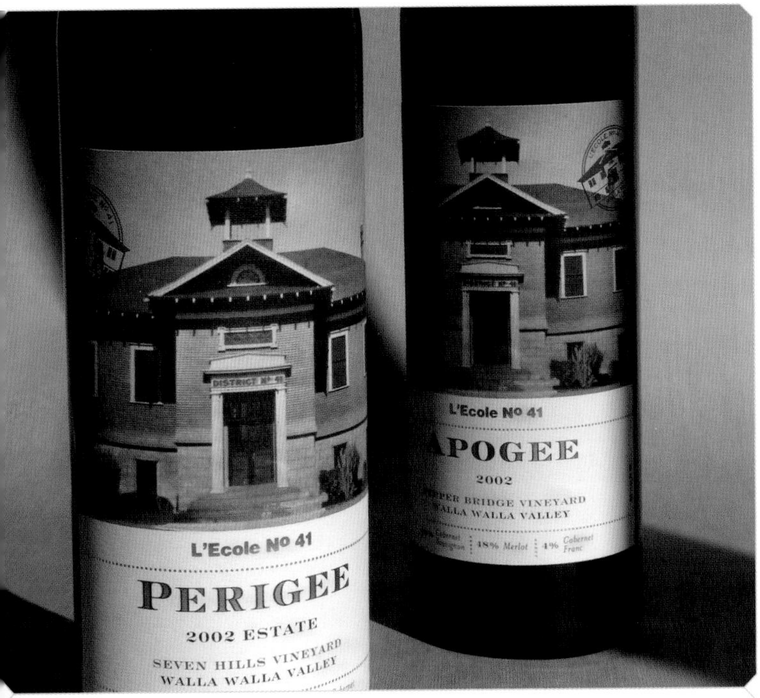

MICHAEL OSBORNE DESIGN
USA
0967

MUCCA DESIGN CORP.
USA
0968

MUCCA DESIGN CORP.
USA
0969

GRETEMAN GROUP
USA
0970

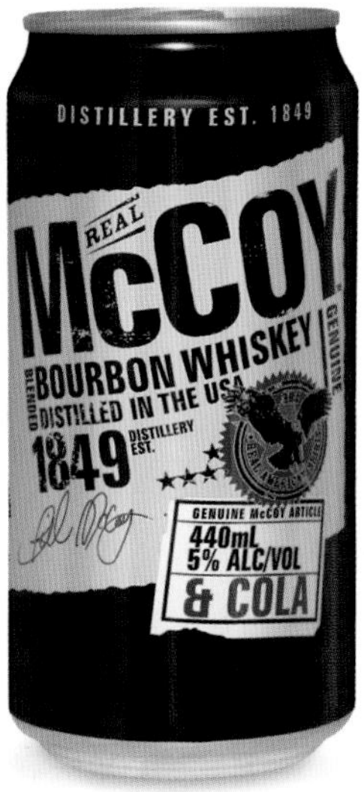

THE CREATIVE METHOD
AUSTRALIA
0971

THE CREATIVE METHOD
AUSTRALIA
0972

THE CREATIVE METHOD
AUSTRALIA
0973

THE CREATIVE METHOD
AUSTRALIA
0974

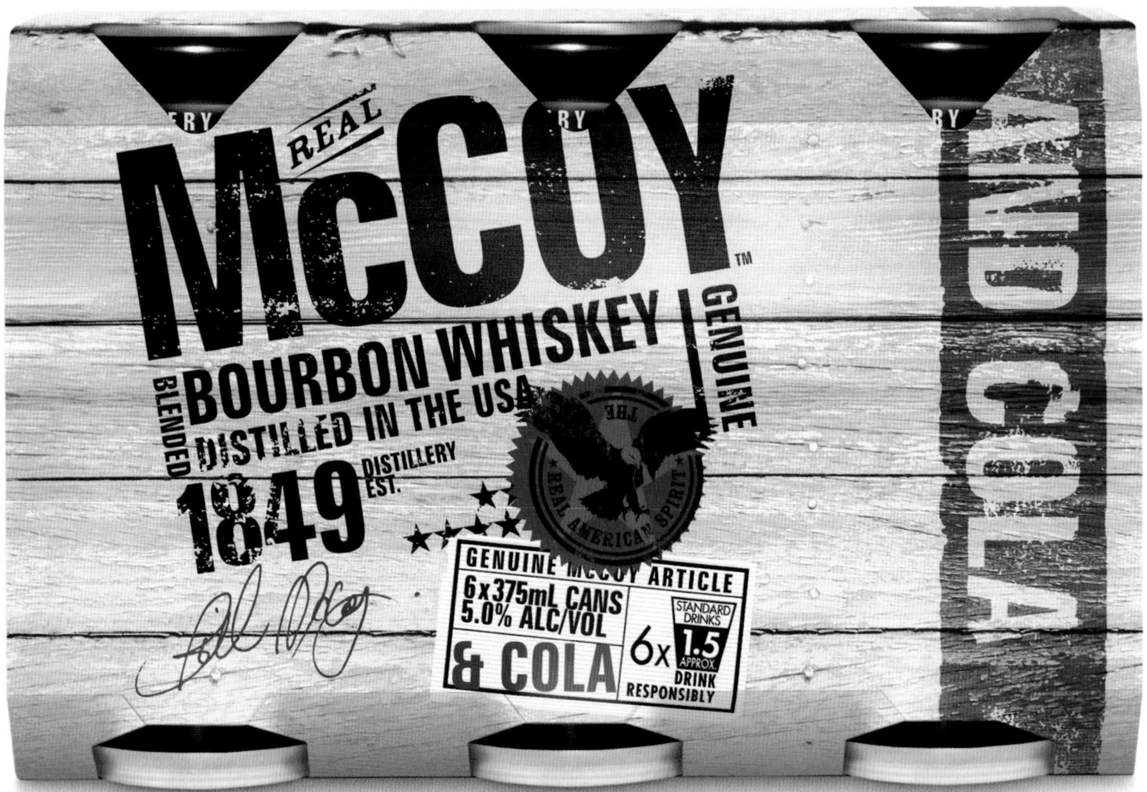
REAL
McCOY
TM
BOURBON WHISKEY
BLENDED
DISTILLED IN THE USA
GENUINE
1849
DISTILLERY
EST.
THE
REAL AMERICAN SPIRIT
GENUINE McCOY ARTICLE
6 x 375mL CANS
5.0% ALC/VOL
& COLA
6x
STANDARD DRINKS
1.5
APPROX.
DRINK RESPONSIBLY
AND COLA

THE CREATIVE METHOD
AUSTRALIA

0975

rick's picks™
phat
beets™
pickled beets
NET WEIGHT: 15 OZ. (427G)

RICK'S PICKS
USA

0976

RICK'S PICKS
USA
0977

RICK'S PICKS
USA
0978

ELEMENT
USA
0979

ID29
USA
0980

ID29
USA
0981

ID29
USA
0982

BLACK EYE DESIGN
CANADA
0983

CORNERSTONE STRATEGIC BRANDING
USA
0984

ID29
USA
0985

CREATIVE COMMUNE / URBAN ACCENTS
USA
0986

GIORGIO DAVANZO DESIGN
USA
0987

CREATIVE COMMUNE / URBAN ACCENTS
USA
0988

CREATIVE COMMUNE / URBAN ACCENTS
USA
0989

CREATIVE COMMUNE / URBAN ACCENTS
USA
0990

CREATIVE COMMUNE / URBAN ACCENTS
USA
0991

CREATIVE COMMUNE / URBAN ACCENTS
USA
0992

CREATIVE COMMUNE / URBAN ACCENTS
USA
0993

CREATIVE COMMUNE / URBAN ACCENTS
USA
0994

URBAN ACCENTS
GOURMET GOBBLER™ KIT
easy, perfect, juicy turkey every time
1
Heavy Duty
Brining Bag
[for Turkeys up to 20lb.]
just add turkey!
[2.5oz | 70g]
Smoky Peppercorn Rub
[16oz | 453g]
Spiced Brine Blend
Nutrition Facts

CREATIVE COMMUNE / URBAN ACCENTS
USA

0995

The Fine Cheese Co.
NATURAL CRACKERS
A delicate and buttery cracker for soft & mild cheeses
150g ℮ net.wt 5.3oz

The Fine Cheese Co.
CHARCOAL CRACKERS
Stylish black crackers suitable for all types of cheese
150g ℮ net.wt 5.3oz

The Fine Cheese Co.
WHOLEMEAL
SPECIALLY MILLED
FLOUR
IN ENGLAND
STONEGROUND
25 kg
WHOLEMEAL CRACKERS
A wheaty and crunchy cracker for strongly flavoured cheeses
150g ℮ net.wt 5.3oz

IRVING
UK

0996

TURNER DUCKWORTH
USA

0997

SONSOLES LLORENS
SPAIN

0998

IRVING
UK

0999

TURNER DUCKWORTH
USA

1000

index

A

212-BIG-BOLT
USA
212-244-2658
woudt@bigbolt.com

0036 – 0040, 0107, 0108, 0302, 0673
Art Director: Pieter Woudt
Designer: Pieter Woudt
Client: Kikkerland

0301
Art Director: Pieter Woudt
Designers: Pieter Woudt, Viktor Jondal
Client: Kikkerland

28 LIMITED BRAND
Germany
+49254-916095-1
mail@no28.de

0718
Art Director: Mirco Kurth
Client: Mactastic

ABSOLUTE ZERO DEGREES
UK
+44 (0) 2077376767
info@absolutezerodegrees.com

0471 – 0475, 0477
Art Director: Keith Stephenson
Designer: Absolute Zero Degrees, UK (with original pattern design by DPA, France)
Client: Pout

0476
Art Director: Keith Stephenson
Designer: Absolute Zero Degrees
Client: Pout

ADDIS CRESON
USA
510-704-7500
jooeun.lee@addiscreson.com

0571
Art Director: John Creson
Designer: Dan Chau
Client: Think Products

ALEXANDER ISLEY INC.
USA
203-544-9692
info@alexanderisley.com

0480, 0481
Art Director: Alexander Isley
Designers: Tara Benyei, Cherith Victorino
Client: Elizabeth Arden Red Door Spas

ALICE CHOCOLATE LLC
USA
917-770-3766
info@alicechocolate.com

0171
Art Directors: Michael Falber, Steven Mark Klein
Client: Alice Chocolate LLC

A L M PROJECT
USA
323-570-0571
andrea@instantdays.com

0300, 0305
Art Director: Andrea Lenardin
Designer: Andrea Lenardin
Client: Sprinkles Cupcakes

AMERICAN APPAREL
USA
213-488-0226
benno@americanapparel.net

0110
Art Director: Benno Russell
Designers: Louise Paradis, Chris Beroine
Client: American Apparel

ANTHEM WORLDWIDE
USA
415-896-9399
heidi.grunenwald@anthemworldwide.com

0201 – 0209
Art Directors: Ron Vandenberg, Brian Lovell
Designer: Michael D. Johnson
Client: Safeway Inc.

ANTIPODES WATER CO.
New Zealand
+64 9 846 9651
andrew@antipodes.co.nz

0535
Art Director: Len Cheeseman
Designer: Len Cheeseman
Client: Antipodes

ANTONIN FERLA
UK
+44 7757 567 603
antonin@antoninferla.com

0373
Art Director: Antonin Ferla
Designers: Antonin Ferla, Aude Crausaz

0417
Art Director: Antonin Ferla
Designer: Antonin Ferla
Client: Antonin Ferla

ARCHRIVAL
USA
402-435-2525
carey@archrival.com

0479, 0482, 0487 – 0495
Art Director: Charles Hull
Designer: Joel Kreutzer
Client: Walgreens

ARLO
USA
773-342-2756
ryan@arlo-tm.com

0062
Art Director: Ryan Halvorsen
Designer: Ryan Halvorsen
Client: Effen Vodka

AUDE FERLA
UK
077 58834637
aude@loremipsum.li

0413, 0440
Designer: Aude Ferla

AURORA DESIGN
UK
+44 628 521234
fusion@aurora-design.co.uk

0118
Art Director: Paul Danbury
Designer: Paul Danbury
Client: Aurora Design

B

BAD STUDIO
USA
404-881-1977, 404-307-8206
scott@badgraphics.com

0922, 0923, 0925, 0949, 0950, 0955
Designer: Scott Banks
Client: Cartoon Network/Adult Swim

BAKKEN CREATIVE CO.
USA
415-225-6036
mbakken@bakkencreativeco.com

0245
Art Director: Michelle Bakken
Designers: Michelle Bakken, Tae Hatayama
Client: Bakken Creative Co.

0409 – 0411
Art Director: Michelle Bakken
Designers: Gina Mondello, Tae Hatayama
Client: Bakken Creative Co.

0578
Art Director: Michelle Bakken
Designers: Gina Mondello, Tae Hatayama
Client: Auberge Resorts

0600
Art Director: Michelle Bakken
Designer: Michelle Bakken
Client: Sumbody

0796
Art Director: Michelle Bakken
Designers: Michelle Bakken, Tae Hatayama
Client: Agustin Huneeus

0881
Art Director: Michelle Bakken
Designer: Tae Hatayama
Client: Staffoon Terje

BBK STUDIO
USA
616-459-4444
Jennifer@bbkstudio.com

0253
Art Director: Yang Kim
Designer: Tim Calkins
Client: BBK Studio

BEARBEAR CREATIVE
USA
714-306-7653
bearbearcreative@gmail.com

0428
Art Director: John Drew
Designer: Sharon Huang
Client: Sequoia National Park

BFG
USA
843-837-9115
leon@bfgcom.com

0111
Art Director: Surasak Lueng
Creative Director: Scott Seymour
Client: BFG

0406 – 0408
Designers: Rachel Schroeder, Jen Young
Client: RJ Reynolds

BLACKBOOKS
USA
954-296-1675
books@blackbookstencils.com

0379
Art Director: Books IIII
Designer: Black Books
Client: Self

BLACK EYE DESIGN
Canada
514-940-2121
jennifer@blackeye.com

0853
Art Director: Michel Vrana
Designer: Grace Cheong
Client: Black Eye

0983
Art Director: Michel Vrana
Designer: Amandine Maillarbauy
Client: Black Eye

BLISS LANE STUDIOS
USA
912-655-6744
leigh@blisslane.com

0803 – 0806
Art Director: Leigh Thompson
Designer: Leigh Thompson
Client: Zhena's Gypsy Tea

BLOOMSBERRY & CO.
USA
978-745-9100
pruett@bloomsberry.com

0307 – 0320

BLUE LOUNGE DESIGN
USA
626-564-2802
Melissa@bluelounge.com

0611
Art Director: Melissa Sunjaya
Designer: Diana Sopha
Client: Trina Turk

0636 – 0645
Art Directors: Melissa Sunjaya, Dominic Symons
Designers: Diana Sopha, Doris Kao, Malek Chalaby
Client: Hint, Inc. – San Francisco

0705
Art Director: Melissa Sunjaya
Designer: Diana Sopha
Client: Betanix

0737
Art Director: Dominic Symons
Designer: Doris Kao
Client: Wing Hing Food

BLUE Q
USA
413-442-1600
Mitch@BlueQ.com

0247 – 0250, 0276 – 0278, 0321 – 0325
Art Director: Mitch Nash
Designer: Haley Johnson
Client: Blue Q

0303, 0324
Art Director: Mitch Nash
Designer: Agnies 2ka Gaspar ska
Client: Blue Q

BONBON LONDON
UK
+44 7932 008 225
mark@bonbonlondon.com

0433, 0434
Art Director: Mark Harper
Designer: Mark Harper
Client: "Smiths" of Smithfield

0814
Art Directors: Mark Harper, Sasha Castling
Designer: Mark Harper
Client: Boardman Bikes

BOY BASTIAENS
The Netherlands
+31 (0)43 3615252
boy@stormhand.com

0377
Art Director: Boy Bastiaens
Designer: Boy Bastiaens
Client: CONFLICT / designmarket

0486
Art Director: Boy Bastiaens
Designer: Boy Bastiaens
Client: Atelier LaDurance

BUCHANAN DESIGN
USA
858-450-1150
bobby@buchanandesign.com

0195
Art Director: Bobby Buchanan
Designers: Bobby Buchanan, Jeanie Nelson, Ryan Skinner, Patrick Faulk
Client: Dolgen Family

0511, 0512
Art Director: Bobby Buchanan
Designers: Bobby Buchanan, Jeanie Nelson
Client: Dolgen Family

0606, 0607
Art Director: Bobby Buchanan
Designers: Bobby Buchanan, Jeanie Nelson, Nick Cottrel, Emily Jansen
Client: Veterinary Specialty Hospital

BUNGALOW CREATIVE
USA
816-421-8770
Christopher@bungalowcreative.com

0518
Art Director: Christopher Huelshorst
Designer: Carrie Kish
Client: Leukemia & Lymphoma Society

0519, 0676
Art Director: Christopher Huelshorst
Designer: Carrie Kish
Client: Bungalow Creative

0648
Art Director: Christopher Huelshorst
Designer: Carrie Kish
Client: Madres Kitchen

0749
Art Director: Christopher Huelshorst
Designer: Carrie Kish
Client: Mildreds Coffeehouse

BURGEFF CO.
Mexico
52 55 55545931
diseno@burgeff.com

0906
Art Director: Patrick Burgeff
Designer: Patrick Burgeff
Client: La Tradicion de la P.

0939
Art Director: Patrick Burgeff
Designer: Patrick Burgeff
Client: El 6Lobo

BURWELL IND. INC.
USA
303-730-3133
marilyn.burwell@gmail.com

0456 – 0466
Art Director: Margot Elena
Designer: Margot Elena

C

CALIENTE CREATIVE
USA
512-627-2607
info@calientecreative.com

0516
Art Directors: Nikki & Nick Lo Bue
Designers: Nikki & Nick Lo Bue
Client: Nikki & Nick Lo Bue

CATALYST STUDIOS
USA
612-339-0735
design@catalyststudios.com

0579
Art Director: Jason Rysavy
Designer: Jon Eckes
Client: Target

0755
Art Director: Jason Rysavy
Designers: Jodi Eckes, Sam Soulek
Client: Snap Industries

0812
Art Director: Shannon Pettini
Directors: Jennifer O'Brien, Sarah Labeineic
Client: Paper Salon

CATO PURNELL
Australia
61 (3) 9419 5566
banki@cato.com.au

0776
Art Director: Cato Purnell
Designer: Cato Purnell
Client: Foster's Group Limited

CAZEE DESIGN
USA
765-532-3746
cazeedesign@yahoo.com

0290
Art Director: Sarah Cazee
Designer: Sarah Cazee
Client: Good Clean Fun

0907
Art Director: Sarah Cazee
Designer: Sarah Cazee
Client: Personal Project

CECI NEW YORK
USA
212-989-0695
info@cecinewyork.com

0506, 0508
Art Director: Lisa Hoffman
Client: Catherine + Jeff

0507, 0514
Art Director: Lisa Hoffman
Client: Julie + Steven

0509
Art Director: Lisa Hoffman
Client: Karen + Andrew

0510, 0517
Art Director: Lisa Hoffman
Client: Angelina Anissimova

0515
Art Director: Lisa Hoffman
Client: Neha + Boris

CHEN DESIGN ASSOCIATES
USA
415-896-5338
info@chendesign.com

0374
Art Director: Joshua C. Chen
Designer: Max Spector
Client: Cordis

CHRONICLE BOOKS
USA
415-537-4373
jenna.cushner@chroniclebooks.com

0001
Art Director: Anne Donnard
Designer: Design Army

0291, 0293
Art Director: Alethea Morrison
Designer: Alethea Morrison
Client: Chronicle Books

0292, 0294
Art Director: Paper Patisserie
Designer: Paper Patisserie
Client: Chronicle Books

0351
Art Director: Catherine Head
Designer: Catherine Head

0352
Art Director: Kristen Hewitt
Designer: Kristen Hewitt
Client: Chronicle Books

0369
Art Director: Anne Donnard
Designers: Alissa Faden, Ellen Lupton

0901
Art Director: Michael Morris
Designer: Michael Morris
Client: Chronicle Books

0902
Art Director: Alethea Morrison
Designer: Deb Wood
Client: Chronicle Books

0903
Art Director: Sara Schneider
Designer: Sara Schneider
Client: Chronicle Books

0904
Art Director: Michael Morris
Designer: Rise and Shine Studio
Client: Chronicle Books

0905
Art Director: Amy E. Achaibou
Designer: Amy E. Achaibou

COMPASS DESIGN
USA
612-339-1599
mitch@compassdesigninc.com

0414, 0429
Art Director: Mitchell Lindgren
Designer: Tom Arthur
Client: August Schell Brewing

CONCRETE DESIGN COMMUNICATIONS
Canada
416-534-9960
mail@concrete.ca

0002 – 0004
Art Directors: Dita Katona, John Pylypczak
Designers: Melissa Agostina, Claire Dawson
Client: For the Dogs

0021 – 0027, 0030, 0370
Art Directors: Diti Katona, John Pylypczak
Designers: Natalie Do, Megan Hunt, Agnes Wong
Client: Susanne Lang Parfumerie

0121 – 0123, 0125, 0169, 0170
Art Directors: Diti Katona, John Pylypczak
Designer: Natalie Do
Client: Susanne Lang Parfumerie

0152, 0153, 0160
Art Directors: Diti Katona, John Pylypczak
Designer: Tom Koukodimos
Client: Pizza Nova

0281, 0282
Art Directors: Diti Katona, John Pylypczak
Designers: Melissa Agostina, Claire Dawson
Client: For the Dogs

CONOVER
USA
619-238-1999
ny@studioconover.com

0376, 0738
Art Director: David Conover
Designer: Nate Yates
Client: Thompson Building Materials

0439
Art Director: David Conover
Designer: Bruno Correia
Client: Self

0782
Art Director: David Conover
Designer: Andrew Salituri
Client: Syndecrete

CORNERSTONE STRATEGIC BRANDING
USA
212-686-6046
keith@cornerstonebranding.com

0427
Client: Boston Beer Co.

0984
Client: Swedish Match

CRASH
USA
847-818-1020
crashcandles@wowway.com

0505
Art Director: Ebony Snow Chafey (of Snow & Graham)
Designer: Becky Corzilius
Client: Crash

CRAVE INC.
USA
561-417-0780
maria@cravebrands.com

0631, 0632
Art Director: David Edmundson
Designers: Russ Martin, Laura Andrews
Client: The Sugar Plum Fairy Baking Co.

0633
Art Director: David Edmundson
Designer: David Edmundson
Client: The Sugar Plum Fairy Baking Co.

0634, 0635
Art Director: David Edmundson
Designer: David Edmundson
Client: Organic Cottage

0671
Art Director: David Edmundson
Designer: David Edmundson
Client: Atlantis Foods

0713
Art Director: David Edmundson
Designer: James Garvin
Client: Fine Tune Golf

0714
Art Director: David Edmundson
Designer: David Edmundson

0719
Art Director: David Edmundson
Designer: Laura Andrews
Client: Delight

0764
Art Director: David Edmundson
Designer: Russ Martin
Client: IQ Beverage Group

CREATIVE COMMUNE/ URBAN ACCENTS
USA
773-244-9100
lkamerad@urbanaccents.com

0660 – 0662, 0722 – 0728, 0739, 0910, 0911, 0986, 0988, 0995
Art Director: Jim Dygas
Designer: Jim Dygas
Client: Urban Accents

0729, 0989 – 0994
Art Director: Jim Dygas
Designer: Annika Shultz
Client: Urban Accents

D

DAVID CARTER DESIGN ASSOCIATES
USA
214-826-4631
ashley@dcadesign.com

0560, 0561
Art Director: Donna Aldridge
Designer: Donna Aldridge
Client: Lajitas Resort/Spa

0934
Art Director: Stephanie Burt
Designer: Stephanie Burt
Client: Lajitas Resort/Spa

DECARLO DESIGN CO.
USA
773-315-0682
nick3141@comcast.net

0947
Art Director: Nick DeCarlo
Designer: Nick DeCarlo
Client: Halfway Jane

DECKER DESIGN
USA
212-633-8588
shannonh@deckerdesign.com

0120
Designer: Lynda Decker
Client: Digital Color Concepts

DELTA ENTERTAINMENT CORPORATION
USA
310-259-8294
michelle@deltamusic.com, justice4me@aol.com

0665 – 0668
Art Director: Michelle Justice
Designer: Pascale Bergara
Client: Delta Entertainment Corporation

DEPERSICO CREATIVE GROUP
USA
610-789-4400
scott@depersico.com

0740
Art Director: Steven Green
Designer: Steven Green
Client: George Weston Bakeries, Inc.

0741
Art Director: Jeff Marshall
Designer: Jeff Marshall
Client: Her Foods, Inc.

DESIGN AHEAD
Germany
0044 201 842060
decker@design-ahead.com

0028
Art Director: Vocker Feddecl
Designer: Nicole Bremenfew
Client: Kaufhof AG

0029
Designer: Theo Decker
Client: Kaufhof AG

0623, 0624
Art Director: Axel Voss
Designer: Axel Voss
Client: Trink Gut

0715
Art Director: Axel Voss
Designer: Axel Voss
Client: Stollwerck

DESIGN BRIDGE
UK
+44 207 814 1182
kirsty.rafferty@designbridge.com

0063
Art Director: Antonia Hayward
Designers: Nicky Triggs, Laurent Robin-Prevaltt
Client: Altia

0065
Art Directors: Antonia Hayward, Daniela Nunzi
Designers: Mimi Ranian, Asa Cook
Client: Interbrew

0353 – 0355
Art Director: Steve Elliott
Designer: Ian Burren
Client: Rieber + Son

0525
Art Director: Graham Shearsby
Designers: Emma Warner, Daniela Nunzi
Client: Constellation Wines

0815
Art Director: Graham Shearsby
Designers: Claire Dale, John Hall
Client: Pernod Ricard

0816
Art Director: Dave Annetts
Designer: William Parr
Client: Asia Pacific Breweries

0845
Art Director: Graham Shearsby
Designers: Daniela Nunzi, Mimi Ranian
Client: Marques de Valoveza

DESIGN FORCE, INC.
USA
856-810-2277
tmininni@designforceinc.com

0675
Art Director: Ted Mininni
Designers: Ted Mininni, Jeremy Doroshon
Client: W.L. Gore & Associates, Inc.

E

ELEMENT
USA
614-447-0906
Jeremy@elementville.com

0283 – 0285
Art Director: Megan Graham
Client: Jeni's Ice Creams

0893 – 0896, 0979
Art Directors: Jeremy Slagle, Megan Graham, John McCollum
Client: Element

0915 – 0920
Art Directors: Jeremy Slagle, John McCollum
Client: Columbus Brewing Co.

ELEVATOR
Croatia
385 98 434 556
tony@elevator.hr

0403 – 0405
Art Director: Tony Adamic
Designers: Lana Vitas, Tony Adamic
Client: Ulola

EMMI
UK
+44 77 5200 1311
hello@emmi.co.uk

0357, 0360, 0420
Art Director: Emmi Salonen
Designer: Emmi Salonen
Client: Self-Promotional

0419
Art Director: Emmi Salonen
Designer: Emmi Salonen
Client: Chieko Mori

EPOS, INC.
USA
310-581-2418
mariko@eposinc.com

0520
Art Director: Barbra Streisand
Designer: Pal Chestnut
Client: Rhino Entertainment Company

0948
Art Director: Gabrielle Raumberger
Designer: Eric Martinez
Client: Spiral Subwave Records International

EVENSON DESIGN GROUP
USA
310-204-1995
mhenry@evensondesign.com

0580
Art Director: Stan Evenson
Designer: Dallas Duncan
Client: EDG

0716
Art Director: Stan Evenson
Designers: Mark Sojka, Weina Dinata
Client: Vuryte

0744
Art Director: Stan Evenson
Designer: Mark Sojka
Client: Blake Brand Growers

0784
Art Director: Stan Evenson
Designer: Kera Scott
Client: Rocomojo

F

FIFTY STRATEGY & CREATIVE
Canada
613-729-0050
Candace@fifty-sc.com

0012 – 0017, 0020
Art Director: Candace Ellicott
Designer: Asami Tanaka
Client: Tai Foong

0758, 0759
Art Director: Candace Ellicott
Designer: Nancy Albert
Client: Soyarie

FIREBELLY DESIGN
USA
773-489-3200
info@firebellydesign.com

0416
Art Director: Dawn Hancock
Designers: Dawn Hancock, Aaron Shimer
Client: Firebelly Design

FLIGHT 001
USA
212-691-1001
john@flight001.com

0031, 0034
Art Directors: John Sencion, Rose Loveras
Designers: Katie Mullins, Anna Tes
Client: Flight 001

0032
Art Director: John Sencion
Designer: Flight 001 Travel Innovations
Client: Flight 001

0033, 0035
Art Directors: John Sencion, Rose Loveras
Designers: Katie Mullins, Anna Tes, Susan Moon Lee, Kayako Ogawa
Client: Flight 001

FRED / EASY ACES INC.
USA
401-333-1311
jason@easyaces.com

0298
Art Director: Jason Amendolara
Designer: Jason Amendolara
Client: Atypyk / Fred

0299, 0329, 0333, 0334, 0336, 0337, 0339, 0344
Art Director: Jason Amendolara
Designer: Christina Sciullo
Client: Fred

0326, 0330 – 0332, 0335
Art Director: Jason Amendolara
Designer: Anthony Petrie
Client: Fred

0327 – 0338, 0340, 0345
Art Director: Jason Amendolara
Designer: Jason Amendolara
Client: Fred

0341
Art Director: Jason Amendolara
Designer: Carl Mitsch
Client: Fred

0342
Art Director: Jason Amendolara
Designer: Luke Boggia
Client: Fred

0343
Art Director: Jason Amendolara
Designers: Jason Amendolara, Anthony Petrie
Client: Fred

FRIENDSWITHYOU
USA
305-519-0873
info@friendswithyou.com

0187 – 0190
Art Director: FriendsWithYou
Designer: FriendsWithYou
Client: FriendsWithYou

FULLBLASTINC.COM
USA
503-227-2002
design@fullblastinc.com

0677
Art Director: Todd Skiles
Designer: Todd Skiles
Client: Microsoft

0797, 0798
Art Director: Todd Skiles
Designer: Todd Skiles
Client: Kozi

FUNNEL : ERIC KASS : UTILITARIAN + COMMERCIAL + FINE : ART
USA
317-590-5355
eric@funnel.tv

0186
Copy: Eric Kass
Designer: Eric Kass
Client: Funnel Paper Goods

0496
Art Director:
Designer: Eric Kass
Client: Films by Francesco

0653
Copy: Eric Kass/Janet Harris
Designer: Eric Kass
Client: The Original Pot-Tee Prize Grab Bag

0940
Copy: Eric Kass/Dori Boneck
Designer: Eric Kass
Client: Maddy's Organic Meals

G

GIORGIO DAVANZO DESIGN
USA
206-328-5031
Giorgio@davanzodesign.com

0422
Designer: Giorgio Davanzo
Client: Struktur (Precept Brands)

0423
Designer: Giorgio Davanzo
Client: Pescaia (Facelli Winery)

0987
Designer: Giorgio Davanzo
Client: Caramoo

GNOME
USA
765-532-3746
cazeedesign@yahoo.com

0435
Art Directors: Sarah Cazee, Nicolle Kasin
Designers: Sarah Cazee, Nicolle Kasin
Client: Fania

GO BIG
USA
781-444-8515
Chris@gobig.info

0011, 0526, 0529
Art Director: Chris Carlozzi
Designer: Chris Carlozzi
Client: Ordonez Wines

0452
Art Director: Gareth Debenham
Designer: Gareth Debenham
Client: Ordonez Wines

0750
Art Director: Chris Carlozzi
Designer: Chris Carlozzi
Client: Delphi Consumer Electronics

0846, 0847
Art Director: Yuko Inagaki
Designer: Yuko Inagaki
Client: Ordonez Wines

GRETEMAN GROUP
USA
316-263-1004
cfarrow@gretemangroup.com

0970
Art Director: Sonia Greteman
Designer: Garrett Fresh
Client: Greteman Group Self Promotion

GRIP
USA
312-906-8020
info@gripdesign.com

0006 – 0010
Art Director: Kelly Kaminski
Designer: Josh Blaylock
Client: Dundee Dermatology

0175
Art Director: Kelly Kaminski
Designer: Josh Blaylock
Client: Wahl

0539
Art Director: Kelly Kaminski
Designer: Josh Blaylock
Client: Landmark

H

HELENA SEO DESIGN
USA
408-830-0086
info@helenaseo.com

0541, 547 – 0550
Art Director: Helena Seo
Designer: Helena Seo
Client: Ineke, LLC.

HELLO! LUCKY
USA
415-355-0008
sarah@hellolucky.com

0295
Art Director: Eunice Moyle
Designer: Sarah Labieniec
Client: Hello! Lucky

HOMART
USA
949-420-2741
Julia@homart.com

0412
Art Director: Julia Long
Designer: Julia Long

0617
Art Directors: Julia Long, Dave Thomas
Designer: Julia Long

0937, 0938, 0936
Art Director: Julia Long
Designers: Julia Long, Carly Gray

HORNALL ANDERSON DESIGN
USA
206-826-2329
c_arbini@hadw.com

0174
Art Director: Mark Popich
Designers: Mark Popich, Ethan Keller, Andrew Well, Jon Graeff, Rachel Lancaster
Client: T-Mobile

0216, 0217
Art Directors: Jack Anderson, Kathy Saito
Designers: Kathy Saito, Sonja Max, Henry Yiu, Yuri Shvets
Client: Ames International

0424
Art Directors: Larry Anderson, Bruce Stigler
Designers: Jay Hilburn, Vu Nguyen
Client: Widmer Brothers Brewery

0483
Art Directors: Lisa Cerveny, Mary Hermes, Julia LaPine
Designers: Mary Hermes, Jana Nishi, Belinda Bowling, Lauren DiRusso, Elmer de la Cruz
Client: O.C. Tanner

0484, 0485, 0799 – 0801
Art Directors: Lisa Cerveny, Sonja Max
Designers: Sonja Max, Belinda Bowling, Ensi Mofasser, Kathy Saito, Beth Grimm, Julie Jacobson
Client: Tahitian Noni

0678 – 0681
Art Directors: Jack Anderson, Andrew Wicklund
Designers: David Bates, Elmer de la Cruz, Jacob Carter, Peter Anderson, Chris Freed, Andrew Wicklund
Client: Microsoft Corporation

0685
Art Directors: Jack Anderson, David Bates
Designers: David Bates, Beth Grimm, Kathleen Gibson, Jacob Carter
Client: Ten Escapes

0793 – 0795
Art Director: Lisa Cerveny
Designers: Mary Hermes, Holy Craven, Belinda Bowling, Tiffany Place, Mary Chin Hutchison
Client: Benjamin Moore

I

ID29
USA
518-687-0268
doug@id29.com

0980 – 0982, 0985
Art Director: Doug Bartow
Designer: Doug Bartow
Client: Brown's Brewing Co.

IRVING
UK
+44 (0) 20 3178 7052
julian@irvingdesigns.com

0059
Art Director: Julian Roberts
Designers: Julian Roberts, Andrea Maloney
Client: Artisan Biscuits

0196
Art Director: Julian Roberts
Designers: Julian Roberts, Brigette Herrod
Client: Artisan Biscuits

0255, 0256, 0261, 0958
Art Director: Julian Roberts
Designers: Julian Roberts, Lucia Gaggiotti
Client: Carluccio's

0371, 0375
Art Director: Julian Roberts
Designers: Julian Roberts, Andrea Maloney
Client: Carluccio's

0537, 0538
Art Director: Julian Roberts
Designer: Dan Adams
Client: Matches

0570
Art Director: Julian Roberts
Designers: Julian Roberts, Andrea Maloney, James Brown
Client: Artisan Biscuits

0615
Art Director: Julian Roberts
Designers: Dan Adams
Client: Carluccio's

0663
Art Director: Julian Roberts
Designers: Julian Roberts, Lucia Gaggiotti
Client: Artisan Biscuits

0760, 0941, 0996, 0999
Art Director: Julian Roberts
Designer: Julian Roberts
Client: The Fine Cheese Co.

0868, 0956, 0957, 0959, 0960
Art Director: Julian Roberts
Designer: Julian Roberts
Client: Carluccio's

J

J. KENNETH ROTHERMICH
USA
212-614-7041
rothermich@yahoo.com

0381
Art Director: J. Kenneth Rothermich
Client: The Great Shakes

JASON MARKK INC.
USA
310-308-2723
jason@jasonmarkk.com

0401
Art Director: Dan Nguyen
Designers: Chhun Tang, April Laviree
Client: Jason Markk Inc.

JEN CADAM DESIGN
USA
858-272-5970
jcadam@san.rr.com

0229 – 0234
Art Director: Jen Cadam
Designers: Jen Cadam, Jose Serrano
Client: The Honest Kitchen

JOED DESIGN INC.
USA
630-993-0080
erebek@joeddesign.com

0149
Art Director: Ed Rebek
Designer: Michelle Brosseau
Client: Paris Presents Inc.

0385
Art Director: Ed Rebek
Designers: Dave Janilek, Ed Rebek
Client: Unilever HPC USA

JOSHUA BLAYLOCK
USA
312-206-6607
blaylong@gmail.com

0513
Art Director: Joshua Blaylock
Designer: Joshua Blaylock
Client: Blaylock & Brower

JOSHUA L. HUDSON
USA
870-476-2900
JLH@JoshuaLHudson.com

0808
Art Director: Joshua L. Hudson
Designer: Joshua L. Hudson
Client: Organic Stash

0822
Art Director: Joshua L. Hudson
Designer: Joshua L. Hudson
Client: S'enivrer

JULIA JURIGA-LAMUT
Austria
+43 669 39 26 041
Julia@juriga.at

0621, 0622
Art Director: Julia Juriga-Lamut
Client: Quinta da Casaboa

K

KASUBA DESIGN
USA
610-202-2905
terri@kasubadesign.com

0211
Designer: Terri Kasuba
Client: Self-promo

KAYU TAI
USA
586-216-9158
ktai@kayulator.com, kayu.tai@interpublic.com, iuphoto@yahoo.com

0612
Art Director: Kayu Tai
Designer: Kayu Tai
Client: Kayu Tai

KBDA
USA
310-255-0902
kim@kbda.com

0746 – 0748
Art Director: Kim Baer
Designer: Allison Bloss
Client: Michelle's Outrageous Morsels

0809
Art Director: Kim Baer
Designer: Barb Cooper
Client: Paula's Salsas

KENDALL ROSS
USA
206-262-0540
david@kendallross.com

0304
Art Director: David Kendall
Client: Made in Washington Stores

0531
Art Director: David Kendall
Client: Precept Brands

0763
Art Director: David Kendall
Client: Jake's BBQ Sauces

0832
Art Director: David Kendall
Client: Sahale Snacks

0927 – 0929
Art Director: David Kendall
Client: SB Northwest

KEN-TSAI LEE DESIGN STUDIO
Taiwan
886-2-87803656
leekentsai@yahoo.com.tw

0382 – 0384
Art Director: Ken-Tsai Lee
Designer: Ken-Tsai Lee
Client: Fonso Interprise Co., Ltd.

KYM ABRAMS DESIGN
USA
312-654-1005
melissad@kad.com

0064
Art Director: Kym Abrams
Designer: Melissa DePasquale
Client: North Lawndale Employment Network

0709
Art Director: Kym Abrams
Designer: Nancy Rockwood
Client: Beansprouts

L

LA SUPERAGENCIA C
Venezuela
+58 412 925 39 63
tennis@tennisandcoffee.com

0368
Art Director: Luis Montenegro Lafont
Designer: Luis Montenegro Lafont
Clients: Todosantos

LEMLEY DESIGN COMPANY
USA
206-285-6900
cameron@lemleydesign.com

0235
Art Director: David Lemley
Designers: Ensi Mofasser, Coventry Jankowski
Client: Sur La Table

0730 – 0734
Art Director: David Lemley
Designers: Pat Snavely, Amber Hahto, Jeremy Wheat, Coventry Jankowski
Client: Sugoi

0789, 0792
Art Director: David Lemley
Designers: Matthew Loyd, Kim Toda, Coventry Jankowski
Client: Grace's Kitchen

0861 – 0863
Art Director: David Lemley
Designers: Bryan Pieratt, Colleen Gray, David Lemley
Client: Tully's Coffee

LESS RAIN
Japan
+81 3 6277 3340
carsten@lessrain.com

0378
Art Director: Lars Eberle
Designers: Lars Eberle, Carsten Schneider
Client: Less Rain

LIPPINCOTT
USA
212-521-0000
www.lippincott.com

0246
Art Directors: Aline Kim, Peter Chun
Designers: Aline Kim, Peter Chun
Client: The Amazing Food Wine Company

LLC "DESIGN DEPOT"
Russia
+7 455-7393373
o.petrushina@designdepot.ru

0743
Art Director: Bankov Piter
Designer: Fedulova Julia
Client: Siberian Gourman

0745
Art Director: Bankov Piter
Designer: Elena Mappyrova
Client: Korovino

0790
Art Director: Bankov Piter
Designer: Elena Mappirova
Client: Vkys Dnia

LLOYD'S GRAPHIC DESIGN LTD.
New Zealand
lloydgraphics@xtra.co.nz

0843
Art Director: Alexander Lloyd
Designer: Alexander Lloyd
Client: Premoli/Single Serve Packaging

LMS DESIGN INC.
USA
203-831-9100
rshear@lmsdesign.com

0151
Art Director: Richard Spears
Client: Vermont Spirits

0707
Art Director: Pamela Shear
Designer: Lael Porcelli
Client: Johnson + Johnson Global Strategic Design

L'OREAL USA – DIESEL FRAGRANCES
USA
212-984-4605
ulindauor@us.loreal.com

0536, 0546
Art Director: Angelica DelPriore
Client: Diesel

M

MARY HUTCHISON DESIGN LLC
USA
206-407-3460
info@maryhutchisondesign.com

0102, 0167, 0710 – 0712
Art Director: Mary Chin Hutchison
Designer: Mary Chin Hutchison
Client: Innovative Beverage Concepts Inc.

0762
Art Director: Mary Chin Hutchison
Designer: Mary Chin Hutchison
Client: WOW Baking Company LLC

METHANE STUDIOS
USA
678-576-5432
Robert@methanestudios.com

0306
Art Directors: Robert Lee, Mark McDevitt
Designers: Robert Lee, Mark McDevitt
Client: Blue Q

METHOD
USA
415-568-4600
anna@methodhome.com

0141 – 0143, 0145 – 0148, 0271, 0272, 0275
Art Director: Josh Handy
Designer: Josh Handy
Client: Method

0173, 0702, 0703, 0706
Designers: Danny Alexander, Josh Handy
Client: Method

MICHAEL COURTNEY DESIGN, INC.
USA
206-329-8188
landon@michaelcourtneydesign.com

0453, 0454
Art Directors: K. Cramer, D. Burgess, M. Courtney
Designers: Landon Fisher, Debra Burgess
Client: Michael Courtney Design, Inc.

MICHAEL OSBORNE DESIGN
USA
415-255-0125
mo@modsf.com

0067, 0783
Art Director: Michael Osborne
Designers: Alice Koswara, Michael Osborne
Client: William-Sonoma

0069
Art Director: Michael Osborne
Designer: Alice Koswara

0071
Art Director: Michael Osborne
Designers: Cody Dingle, Michael Osborne
Client: Sake2me

0598
Art Director: Michael Osborne
Designer: Sheri Kuniyuki
Client: William-Sonoma

0599
Art Director: Michael Osborne
Designer: Michael Osborne
Client: Target

0613
Art Director: Michael Osborne
Designer: Michael Osborne
Client: Gymboree

0625
Art Director: Michael Osborne
Designers: Beth Leonardo, Michael Osborne
Client: Silverado Vineyards

0699, 0742, 0831
Art Director: Michael Osborne
Designer: Alice Koswara
Client: Kettle Foods

0708
Art Director: Michael Osborne
Designers: Alice Koswara, Michael Osborne
Client: Square One Organic Spirits, LLC

0770
Art Director: Michael Osborne
Designer: Alice Koswara
Client: Brown Foreman

0777
Art Director: Michael Osborne
Designers: Alice Koswara, Michael Osborne
Client: Jax Vineyards

0791, 0865, 0912
Art Director: Michael Osborne
Designer: Alice Koswara
Client: William-Sonoma

0802
Art Director: Michael Osborne
Designer: Beth Leonardo, Michael Osborne
Client: Tamas Estates

0967
Art Director: Michael Osborne
Designer: Michael Osborne
Client: L'Ecole No:41

MICHAEL SCHWAB STUDIO
USA
415-257-5792
studio@michaelschwab.com

0882 – 0890
Designer: Michael Schwab
Client: Larabar

MILCH DESIGN GMBH
Germany
0049-89-520466-0
Judith_may@milch-design.de

0751 – 0754
Art Director: Friedel Patzak
Designer: Friedel Patzak
Client: Elgato Systems

MILLER CREATIVE
USA
732-600-3933
info@yaelmiller.com

0557
Designer: Yael Miller
Client: Astor Chocolate, Marriott Hotels Intl.

0558
Designer: Yael Miller
Client: Astor Chocolate

MILTON GLASER, INC.
USA
212-899-3161
studio@miltonglaser.com

0362 – 0365
Art Director: Milton Glaser
Designer: Milton Glaser
Client: Brooklyn Brewery

0701
Art Director: Milton Glaser
Designers: Deborah Adler, Milton Glaser
Client: Medline Industries

0778
Art Director: Milton Glaser
Designer: Milton Glaser
Client: Drinks America

MIRIELLO GRAFICO
USA
619-234-1124
pronto@miriellografico.com

0720
Art Director: Dennis Garcia
Designers: Josh Higgins, Dennis Garcia
Client: Mondo USA

0813
Art Directors: Craig Barez (TaylorMade), Dennis Garcia
Designers: Dennis Garcia, Robert Palmer (Production)
Client: TaylorMade

MODERN DOG DESIGN CO.
USA
206-789-7667
rob@moderndog.com

0296
Art Director: Vittorio Costarella
Designer: Vittorio Costarella
Client: Blue Q

0297
Art Director: Robert Zwiebel
Designer: Robert Zwiebel
Client: Blue Q

MORLA DESIGN
USA
415-543-7214
arlene@morladesign.com

0168
Art Director: Jennifer Morla
Designer: Jennifer Morla
Client: Discovery Channel

0540
Art Director: Jennifer Morla
Designers: Jennifer Morla, Angela Williams
Client: Shaklee Corporation

0563
Art Director: Jennifer Morla
Designer: Jennifer Morla
Client: Cocolat

MUCCA DESIGN CORP.
USA
212-965-9821
Sylvia.paret@muccadesign.com

0095
Art Director: Matteo Bologna
Designer: Christine Celic Strohl
Client: Washington Square Restaurant

0254
Art Director: Matteo Bologna
Designer: Andrea Brown
Client: Gracious Gourmet

0418
Art Director: Matteo Bologna
Designer: Matteo Bologna
Client: Kartell

0856, 0859
Art Director: Matteo Bologna
Designer: Christine Celic Strohl
Client: Butterfield Market

0857, 0858, 0860, 0966, 0968, 0969
Art Director: Matteo Bologna
Designer: Andrea Brown
Client: Sant Ambroeus Restaurant

0869
Art Director: Matteo Bologna
Designer: Matteo Bologna
Client: Schiller's Restaurant

0870
Art Director: Matteo Bologna
Designer: Christine Celic Strohl
Client: Balthazar Restaurant

MUKU STUDIOS, LLC
USA
407-459-0118
muku@mukustudios.com

0785
Art Director: Jana D. Morgan
Designer: Jana D. Morgan
Client: Self-promotional project

MYINT DESIGN
USA
415-626-0522
marla@myintdesign.com

0144
Art Directors: Karin Myint, Carole Jeung
Designers: Karin Myint, Carole Jeung
Client: Myint Design

0854
Art Director: Karin Myint
Designer: Karin Myint
Client: Pope Creek Ranch, Inc.

N

NIEDERMEIER DESIGN
USA
206-351-3927
kurt@kngraphicdesign.com

0467 – 0470
Art Director: Kurt Niedermeier
Designer: Kurt Niedermeier
Client: Queen Mary Tea

0694 – 0697
Art Director: Kurt Niedermeier
Designer: Kurt Niedermeier
Client: Sahale Snacks

0851, 0852, 0855
Art Director: Kurt Niedermeier
Designer: Kurt Niedermeier
Client: Leatherback Printer

NUDO LTD
UK
+44 (0) 207 617 7235
adopt@nudo-italia.com

0841
Art Director:
Designer: Madeleine Rogers
Client: Nudo LTD

O

O! ADVERTISING & DESIGN
Iceland
+354 864 6640
maria@oid.is

0654 – 0657
Art Director: Maria Ericsoottir
Designer: Maria Ericsoottir
Client: Himnesk Hollusta

0942, 0945
Art Director: Maria Ericsoottir
Designer: Maria Ericsoottir
Client: Mjolka

OBJECT 9
USA
225-268-9899
andy@object9.com

0421
Art Director: Object 9
Designer: Object 9
Client: Diageo / Red Stripe

0425
Art Director: Object 9
Designer: Object 9
Client: Flying Dog Brewery

0842
Art Director: Object 9
Designer: Object 9
Client: Café Bom Dia

OCTAVO DESIGN
Australia
+613 9686 4703
info@octavodesign.com.au

0628, 0629
Art Director: Gary Domoney
Designer: Gary Domoney
Client: Pier 10 Vineyard

0630
Art Director: Gary Domoney
Designer: Gary Domoney
Client: Mornington Estate Winery

OKOLITA M
USA
312-213-1913
mark@okolita.com

0674
Art Director: Mark Okolita
Designer: Mark Okolita
Client: Lava World International

OLD NAVY
USA
415-832-1496
jason_rosenberg@gap.com

0432
Art Director: Jason H. Rosenberg
Designer: Jason H. Rosenberg
Client: Old Navy

ON-PURPOS, INC.
USA
515-699-1666
Austin@onpurpos.com

0761
Art Director: Dan Fish
Designer: Austin Van Lear
Client: Meredith Corporation Brand Licensing

OUTSET, INC.
USA
952-361-0029
sarah.osborn@outsetinc.com

0051 – 0058, 0060
Art Director: Amy Anderson
Designer: Sarah Osborn
Client: Outset, Inc.

OXIDE DESIGN CO.
USA
402-344-0168
drew@oxidedesign.com

0346
Art Director: Drew Davies
Designers: Drew Davies, Joe Sparano
Client: Oxide Design Co.

0347
Art Director: Drew Davies
Designer: Drew Davies
Client: Nebraska AIDS Project

P

P&W
UK
0044 (0) 207 723 8899
lee@p-and-w.com

0018, 0019
Art Directors: Simon Pemberton, Adrian Whitefoord
Designer: Wes Anson
Client: Fresh Pasta Co.

0103 – 0105
Art Directors: Simon Pemberton, Adrian Whitefoord
Designer: Lee Newham
Client: Lyme Regis

0154 – 0159
Art Directors: Simon Pemberton, Adrian Whitefoord
Designer: Wes Anson
Client: Tesco

0197 – 0200, 0218 – 0220, 0693, 0698, 0834 – 0836, 0844
Art Directors: Simon Pemberton, Adrian Whitefoord
Designer: Lee Newham
Client: Tesco

0224, 0225
Art Directors: Simon Pemberton, Adrian Whitefoord
Client: Tesco

0430, 0551, 0555, 0646, 0647, 0651, 0652
Art Directors: Simon Pemberton, Adrian Whitefoord
Designer: Lee Newham
Client: WSI

0552 – 0554
Art Directors: Simon Pemberton, Adrian Whitefoord
Designer: Lee Newham
Client: Hill Station

0649
Art Director: Adrian Whitefoord
Designer: Wes Anson
Client: Fresh & Easy

0780, 0781
Art Directors: Simon Pemberton, Adrian Whitefoord
Designer: Lee Newham
Client: PLT

0828 – 0830
Art Directors: Simon Pemberton, Adrian Whitefoord
Designer: Wes Anson
Client: Starbucks

0837, 0838, 0840
Art Directors: Simon Pemberton, Adrian Whitefoord
Designer: Sue Butroid
Client: Tesco

PANGEA ORGANICS
USA
877-679-5854
libby@pangeaorganics.com

0124, 0126 – 0135, 0138, 0140
Art Director: Joshua S. Onysko
Designer: Ian Groulx
Client: Pangea Organics

0136, 0137, 0139
Art Director: Joshua S. Onysko
Designer: Ian Groulx
Client: Pangea Organics

PAPA, INC.
Croatia
00 385 91 2530235
dubpapa@inet.hr

0183, 0184
Art Director: Dubravko Papa
Designer: Dubravko Papa
Client: Geofoto

PEACE COFFEE
USA
612-870-3440
mel@peacecoffee.com

0952 – 0954
Designer: Haley Johnson
Client: Peace Coffee

PEARLFISHER
UK
011 44 207 603 8666
shaun@pearlfisher,com

0150
Art Director: Jonathan Ford
Designer: Sarah Butler
Client: Nude Cosmetics

0166
Art Director: Shaun Bowen
Designer: Natalie Chung
Client: Dr Stuart's

0191 – 0194
Art Director: Shaun Bowen
Designer: Sarah Pidgeon
Client: Innocence

PEARLFISHER
USA
212-604-0601
lisa@pearlfisher.com

0210
Art Director: Lisa Simpson
Designer: Lisa Simpson
Client: OOPS

0273, 0274
Art Director: Lisa Simpson
Designer: Lisa Simpson
Client: The Hershey Company

0559
Art Director: Lisa Simpson
Designer: Lisa Simpson
Client: Mane, USA

PHILIPPE BECKER DESIGN
USA
415-348-0054
erin@pbdsf.com

0176
Creative Director: Philippe Becker
Designers: Coco Qiu, Melanie Halim
Client: Wally's Food Co.

0564, 0864
Art Director: Philippe Becker
Designers: Melanie Halim, Mariko Muto
Client: Williams-Sonoma

0616
Art Director: Philippe Becker
Designers: Barkha Wadia, Mariko Muto, Coco Qiu
Client: Levlad, Inc.

0650
Art Director: Philippe Becker
Designer: Melanie Halim
Client: Williams-Sonoma

0700
Art Director: Philippe Becker
Designers: Heather Allen, Melanie Halim
Client: Artisan Wine Group

0717
Art Director: Philippe Becker
Designers: Melanie Halim, Coco Qiu, Marcus Carini
Client: Clean Well

0756, 0757
Art Director: Philippe Becker
Designer: Barkha Wadia
Client: Safeway, Inc.

0769
Art Director: Philippe Becker
Designers: Melanie Halim, Jay Cabalquinto, Barkha Wadia
Client: T-Mobile

0872, 0875, 0877, 0878
Art Director: Philippe Becker
Designers: Melanie Halim, Coco Qiu
Client: Williams-Sonoma

0943
Art Director: Philippe Becker
Designers: Barkha Wadia, Coco Qiu
Client: Williams-Sonoma

PINK BLUE BLACK & ORANGE CO., LTD.
Thailand
662-300-5124
mailus@colorparty.com

0114
Art Director: Mr. Punlarp Punnotok
Designer: Ms. Patsuda Rochanaluk
Client: Chamni Studio

0161 – 0163
Art Director: Mr. Siam Attariya
Designer: Mr. Nattapol Poonpiriya
Client: Central Pattana (CPN)

0181
Art Director: Mr. Siam Attariya
Designer: Mr. Siam Attariya
Client: Pink Blue Black & Orange Co., Ltd.

0182, 0185
Art Director: Mr. Vichean Tow
Designer: Mr. Siriwat Sangsomsap
Client: Pink Blue Black & Orange Co., Ltd

0251
Art Director: Mr. Punlarp Punnotok
Designer: Ms. Patsuda Rochanaluk
Client: Galio Thai

0252, 0260
Art Director: Ms. Patsuda Rochanaluk
Designer: Ms. Manisa Lekpratoon
Client: Galio Thai

PRINCIPLE
USA
713-521-1625
Pamela@designbyprinciple.com

0005
Art Director: Pamela Zuccker
Designer: Pamela Zuccker
Client: Paddywax

0556, 0610
Art Director: Pamela Zuccker
Designer: Allyson Lack
Client: Paddywax

0614
Art Director: Pamela Zuccker
Designer: Jennifer Sukis
Client: Paddywax

R

RANDY MOSHER DESIGN
USA
773-973-0240
randymosher@rcn.com

0426
Art Director: Randy Mosher
Designer: Randy Mosher
Client: Aardvark Cider

0817, 0818
Art Director: Randy Mosher
Designer: Randy Mosher
Client: Two Brothers Brewing

0933
Art Director: Randy Mosher
Designer: Randy Mosher
Client: Palisade Brewing Co.

RICK'S PICKS
USA
212-358-0428
jina@rickspicksnyc.com

0976 – 0978
Art Director: Stiletto NYC
Designer: Stiletto NYC
Client: Rick's Picks

ROME & GOLD CREATIVE
USA
505-897-0870
Lorenzo@rgcreative.com

0258, 0259
Art Director: Lorenzo Romero
Designer: Robert E. Goldie
Client: Boba Tea Company

0672
Art Director: Lorenzo Romero
Designer: Robert E. Goldie
Client: Innovasic Semiconductors

RULE29
USA
630-262-1009
Justin@rule29.com

0257
Art Director: Justin Ahrens
Designers: Justin Ahrens, Kerri Liu, Josh Jensen, Dan Hassenplug
Client: O'Neil Printing

S

SAGMEISTER INC.
USA
212-647-1789
info@sagmeister.com

0366, 0415, 0944

SALVARTES DISEÑO Y PUBLICIDAD
Spain
676 757 929
Salva@salvartes.com

0164
Art Director: Salva Garcia-Ripoll Toledano
Designer: Manu Vazquez
Client: Roberto Jean

SATELLITE DESIGN
USA
415-371-1610
amy@satellite-design.com

0735, 0736
Art Director: Amy Gustincic
Designer: Amy Gustincic
Client: The North Face

SATELLITES MISTAKEN FOR STARS
Austria
0043 1 4805167
alex@satellitesmistakenforstars.com

0380
Art Director: Alexander Egger
Designer: Alexander Egger

SAYLES GRAPHIC DESIGN
USA
515-279-2922
sheree@salyesdesign.com

0389
Art Director: John Sayles
Designer: John Sayles
Client: MaDIKwe

0951
Art Director: John Sayles
Designer: John Sayles
Client: RotoRooter

SEED
Singapore
(65) 62265422
mark@seed.us.com

0041 – 0047
Art Director: Mark Walker
Designer: Mark Walker
Client: Wild Bunch

SEGURA INC.
USA
773-862-5667
carlos@5inch.com

0436, 0437
Art Director: Carlos Segura
Designer: Segura Inc.

SHINE ADVERTISING
USA
608-442-7373
kleslie@shinenorth.com

0109
Art Directors: Michael Kriefski, John Krull
Client: Umi Shoes

0924
Art Director: Michael Kriefski
Designer: Chad Bollenbach
Client: Shine Advertising

SIBLEY PETEET DESIGN – DALLAS
USA
214-969-1050
denise@spddallas.com

0773, 0775
Art Director: Don Sibley
Designer: Brandon Kirk
Client: The Gambrinus Company

0774
Art Director: Don Sibley
Designer: Geoff German
Client: The Gambrinus Company

0807
Art Director: Don Sibley
Designer: Rey Latham
Client: Nokia

SOCKEYE CREATIVE
USA
503-226-3843
rwees@sockeyecreative.com

0930 – 0932
Art Director: Peter Metz
Designer: Robert Wees
Client: MacTarnahan's Brewing

SONSOLES LLORENS
Spain
+34 934 124 171
info@sonsoles.com

0106, 0566 – 0569, 0602 – 0605, 0608, 0619, 0670, 0998
Art Director: Sonsoles Llorens
Designer: Sonsoles Llorens
Client: Sans&Sans Finetea Merchants

SPARK STUDIO
Australia
+613 9686 4703
info@sparkstudio.com.au

0626
Art Director: Gary Domoney
Designer: Adam Pugh
Client: Shelmerdine Winery

0627
Art Director: Sean Pethick
Designer: Sean Pethick
Client: Leaning Edge

0689
Art Director: Sean Pethick
Designer: Sean Pethick
Client: National Transport Institute of Australia

0810
Art Director: Sean Pethick
Designer: Natalie Leys
Client: Comforts

0811
Art Director: Sean Pethick
Designer: Sean Pethick
Client: Cottons

STUDIOBENBEN
USA
310-804-8751
studio@benschlitter.com

0074, 0609
Art Director: Ben Schlitter
Designer: Ben Schlitter

SUBPLOT DESIGN INC.
USA
604-685-2990
steph@subplot.com

0391 – 0400, 0961 – 0964
Art Directors: Matthew Clark, Roy White
Designer: Matthew Clark
Client: Fully Loaded Tea

SYNTHETIC INFATUATION
USA
312-203-6267
lucas@synth.tc

0367
Art Directors: Lucas Buick, Ryan Dorshorst
Client: Synthetic

T

TABLE BRIWE INK
USA
845-758-2688
jfaso@twyluzbeverge.com

0066
Art Director: Patricia Spencer
Designer: Patricia Spencer

TEA FORTÉ
USA
978-369-7777
koleary@teaforte.com

0497 – 0500
Art Director: Peter Hewitt
Designer: Peter Hewitt
Client: Tea Forte

HE COLLECTIVE DESIGN ONSULTANTS
ustralia
61 2 9281 5533
owena@thecollectivedesign.com.au

072, 0073, 0075
rt Director:
esigner: Margaret Nolan
lient: Richfield Vineyard

521 – 0523
esigner: Margaret Nolan
lient: Codralook by Yabby Lake
nternational

524
esigner: Margaret Nolan
lient: Dexter Wines

530
esigner: Margaret Nolan
lient: Cape Barren Wines

532
esigner: Margaret Nolan
lient: Cooba East Station

THE CREATIVE METHOD
Australia
0061 2 823/9977
ony@thecreativemethod.com

0356, 0358, 0359
Art Director: Tony Ibbotson
Designer: Tony Ibbotson
Client: Guzman Y Gomez

0455
Art Director: Tony Ibbotson
Designer: Tony Ibbotson
Client: Paul Sibqaa – Grumblebone
Est.

0965, 0971 – 0975
Art Director: Tony Ibbotson
Designer: Tony Ibbotson
Client: Diageo Australia

THE DECODER RING DESIGN CONCERN
USA
512-236-1610
Christian@thedecoderring.com

0372
Designer: Christian Helms
Client: Signature Sounds Records

0431
Art Directors: Isaac Brock, Christian Helms, Naheed Simjee
Designer: Christian Helms
Client: Modest Mouse and Epic/Sony BMG

0946
Designer: Christian Helms
Client: Aaron Goldman Films and Lucero

THE JONES GROUP
USA
404-523-2606
Vicky@thejonesgroup.com

0664
Art Director: Vicky Jones
Designer: Kendra Lively
Client: The Jones Group

0704
Art Director: Vicky Jones
Designer: Kendra Lively
Client: OneSource

THE O GROUP
USA
212-398-0100
krothermich@ogroup.net

0565
Art Director: Jason B. Cohen
Designer: Jason Kirshenblatt
Client: Robert Marc

0819
Art Director: Jason B. Cohen
Designer: J. Kenneth Rothermich
Client: Hennessy Cognac

THE MICRO AGENCY
Canada
416-449-8050
Raymond@themicroagency.com

0771, 0772, 0779
Art Director: Raymond Waters
Designer: Raymond Waters
Client: Great Lakes Brewery – Toronto

THE REPUBLIC OF TEA
USA
415-382-3443
marideth@republicoftea.com

0262 – 0270
Art Director: Gina Amador
Designer: Gina Amador

THYNK DESIGN, INC.
USA
630-323-1020
migreyes@thynk.com

0874
Art Director: Dale Rehus
Designer: Dan Kazmer
Client: First Choice

TILKA DESIGN
USA
612-664-8994
sbusse@tilka.com

0119
Art Director: Tilka Design
Designer: Tilka Design
Client: Tilka Design

TOKY BRANDING + DESIGN
USA
314-534-2000
eric@toky.com

0279, 0280
Art Director: Eric Thoelke
Designer: Eric Thoelke
Client: Bissinger's Handcrafted Chocolatier

0669, 0871, 0876, 0880, 0909
Art Director: Eric Thoelke
Designer: Jamie Banks-George
Client: Eckert's Country Store

0572 – 0574
Art Director: Eric Thoelke
Designers: Jamie Banks-George, Tara Pederson
Client: Bissinger's Handcrafted Chocolatier

0581 – 0586, 0592 – 0595, 0601
Art Director: Eric Thoelke
Designer: Jamie Banks-George
Client: Bissinger's Handcrafted Chocolatier

0587-0590
Art Director: Eric Thoelke
Designer: Jamie Banks-George
Client: Bissinger's Handcrafted Chocolatier

0591
Art Director: Eric Thoelke
Designers: Jamie Banks-George, Geoff Story
Client: Bissinger's Handcrafted Chocolatier

0618
Art Director: Eric Thoelke
Designers: Eric Thoelke, Karin Soukup, Geoff Story
Client: S. King Collection

0833, 0908, 0913, 0914
Art Director: Eric Thoelke
Designer: Jamie Banks-George
Client: The Smokehouse Market

0873, 0879
Art Director: Eric Thoelke
Designer: Jamie Banks-George
Client: Straub's Fine Grocers

TURNER DUCKWORTH
USA
415-675-7777
lisa@turnerduckworth.com

0048 – 0050, 0848 – 0850
Creative Directors: David Turner, Bruce Duckworth
Designer: Christian Eager
Client: Waitrose

0061
Creative Directors: David Turner, Bruce Duckworth
Designers: Shaun Rosenberger, Brittany Hull
Client: Click Wire Group

0070
Creative Directors: David Turner, Bruce Duckworth
Designer: Shaun Rosenberger
Client: Click Wine Group

0076 – 0083
Creative Directors: David Turner, Bruce Duckworth
Designers: Paula Talford, Mike Harris, Charlotte Barres
Client: Homebase Ltd.

0084 – 0090
Creative Directors: David Turner, Bruce Duckworth
Designer: Christian Eager
Client: Homebase Ltd.

0091 – 0094
Creative Directors: David Turner, Bruce Duckworth
Designer: Shaun Rosenberger
Client: Shaklee

0096 – 0101, 0898 – 0900
Creative Directors: David Turner, Bruce Duckworth
Designer: Sarah Moffat
Client: Waitrose

0165
Creative Directors: David Turner, Bruce Duckworth
Designer: Jamie McCathie
Client: Liz Earle

0177, 0178
Creative Directors: David Turner, Bruce Duckworth
Designers: Bruce Duckworth, Janice Davidson
Client: Liz Earle Cosmetics

0179, 0180
Creative Directors: David Turner, Bruce Duckworth
Designers: Sarah Moffat, Chris Garvey, Aaron Maurer
Client: The Coca-Cola Company

0212 – 0215
Creative Directors: David Turner, Bruce Duckworth
Designer: Sam Lachlan
Client: Superdrug

0221 – 0223
Creative Directors: David Turner, Bruce Duckworth
Designers: Mike Harris, Sofie Moller
Client: Homebase Ltd.

0226 – 0228, 0575 – 0577, 0692
Creative Directors: David Turner, Bruce Duckworth
Designer: Sam Lachlan
Client: Waitrose

0240 – 0244
Creative Directors: David Turner, Bruce Duckworth
Designer: Bruce Duckworth
Client: Waitrose

0286 – 0289
Creative Directors: David Turner, Bruce Duckworth
Designers: Sam Lachlan, Christian Eager
Client: Superdrug

0478
Creative Directors: David Turner, Bruce Duckworth
Designer: Chris Garvey
Client: Murad

0501 – 0504
Creative Directors: David Turner, Bruce Duckworth
Designer: Jamie McCathie
Client: Superdrug

0527, 0528
Creative Directors: David Turner, Bruce Duckworth
Designer: Chris Garvey
Client: Click Wine Group

0533, 0534
Creative Directors: David Turner, Bruce Duckworth
Designers: David Turner, Shawn Rosenberger, Chris Garvey, Britany Hull, Rachel Shaw
Client: Click Wine Group

0620
Creative Directors: David Turner, Bruce Duckworth
Designer: Sarah Moffat
Client: Virgin Atlantic

0682 – 0684, 0897, 0997, 1000
Creative Directors: David Turner, Bruce Duckworth
Designer: Jamie McCathie
Client: Waitrose

0686, 0687, 0765 – 0768
Creative Directors: David Turner, Bruce Duckworth
Designers: Shawn Rosenberger, Ann Jordan, Josh Michels, Rebecca Williams, Brittany Hull, Radu Ranga
Client: Motorola

0690, 0691
Creative Directors: David Turner, Bruce Duckworth
Designer: Sarah Moffat
Client: Coca-Cola GB

0823 – 0827
Creative Directors: David Turner, Bruce Duckworth
Designers: Shawn Rosenberger, David Turner
Client: Method

0839
Creative Directors: David Turner, Bruce Duckworth
Designer: Chris Garvey
Client: Popchips

0866, 0867
Creative Directors: David Turner, Bruce Duckworth
Designer: David Turner
Client: McKenzie River Corp.

0891, 0892
Creative Directors: David Turner, Bruce Duckworth
Designers: Bruce Duckworth, Bob Celez

0935
Creative Directors: David Turner, Bruce Duckworth
Designer: Brittany Hull
Client: Tommy's Margaritas

TURNSTYLE
USA
206-297-7350
info@turnstylestudio.com

0068
Art Director: Steve Watson
Designer: Steve Watson
Client: DRY Soda Company

0172
Art Director: Steve Watson
Designer: Steve Watson
Client: Lello

0236
Art Director: Steve Watson
Designer: Steve Watson
Client: Full Tank Foods

0386, 0387, 0390
Art Director: Steve Watson
Designer: Jason Gomez
Client: ReelWorld

U

UNIVERSITY OF NEBRASKA AT KEARNEY
USA
308-865-8353
gabere@gmail.com

0562
Art Director: Wuthichai Choonhasakulchoke
Designer: Gabriello Re
Client: Arcadia organic teas

W

WALLACE CHURCH, INC.
USA
212-755-2903
calley@wallacechurch.com

0348 – 0350
Art Director: Stan Church
Designers: Nin Glaister, Jhomy Irrazabai, Jodi Lubrich
Client: Ciao Bella Gelato Co., Inc.

0402
Art Directors: Stan Church, John Bruno
Designer: Akira Yasuda
Client: The Dial Corporation

0596, 0597
Art Director: Stan Church
Designers: Lawrence Haggerty, Jhomy Irrazabai
Client: Pepperidge Farm

0658
Art Director: Stan Church
Designer: Claire Reece-Raybould
Client: American Italian Pasta Company

0659
Art Director: Stan Church
Designer: Stan Church

0688
Art Director: Stan Church
Designer: John Bruno
Client: The Gillette Company

0721, 0921
Art Director: Stan Church
Designers: (Final) Allen Ghorian, (Concept) Jhomy Irrazabai
Client: Bradford Soaps

0820
Art Directors: Stan Church, Lawrence Haggerty
Designer: Camilla Kristiansen
Client: Wyattzier

0821
Art Director: Stan Church
Designer: Lawrence Haggerty
Client: Wyattzier

WHITERHINO CREATIVE
Australia
613 9428 8896
peter@whiterhino.com.au

0926

WOLKEN COMMUNICA
USA
206-545-1696
kurt@wolkencommunica.com

0438
Art Director: Kurt Wolken
Designer: Johann Gomez
Client: True Benefits

Y

Y STUDIOS LLC
USA
415-206-0622
y@ystudios.com

0441 – 0451
Art Directors: Y Studios: Wai-loong Lim/Mudhaus: Michael Rutchik
Designers: Y Studios: Wai-loong Lim/ Mudhaus: Michael Rutchik
Client: Presidio Brands

YELLOBEE STUDIO
USA
404-249-6407
ascheel@yellobee.com

0786 – 0788
Art Director: Alison Scheel
Designer: Rhonda Dennis
Client: Via Elisa

YIYING LU DESIGN
Australia
61 2 9280 1418
helloyiying@gmail.com

0237 – 0239
Art Director: Yiying Lu
Designer: Yiying Lu
Client: JWT Shanghai

Z

ZION GRAPHICS
Sweden
+46 8 644 37 58
ricky@ziongraphics.com

0112, 0113
Art Director: Ricky Tillblad
Designer: Ricky Tillblad
Client: La'Mode

0115 – 0117
Art Director: Ricky Tillblad
Designer: Ricky Tillblad
Client: Size Records

0361, 0388
Art Director: Ricky Tillblad
Designer: Ricky Tillblad
Client: EMI Music